Maxwell Street

Maxwell Street

WRITING AND THINKING PLACE

Tim Cresswell

University of Chicago Press   Chicago and London

The University of Chicago Press, Chicago 60637
The University of Chicago Press, Ltd., London

Published 2019
Printed in the United States of America

28 27 26 25 24 23 22 21 20 19   1 2 3 4 5

ISBN-13: 978-0-226-60411-4 (cloth)
ISBN-13: 978-0-226-60425-1 (paper)
ISBN-13: 978-0-226-60439-8 (e-book)
DOI: https://doi.org/10.7208/chicago/9780226604398.001.0001

Library of Congress Cataloging-in-Publication Data

Library of Congress Cataloging-in-Publication Data

Names: Cresswell, Tim, author.
Title: Maxwell Street : writing and thinking place / Tim Cresswell.
Description: Chicago ; London : University of Chicago Press, 2019. | Includes bibliographical references and index.
Identifiers: LCCN 2018049441 | ISBN 9780226604114 (cloth : alk. paper) | ISBN 9780226604251 (pbk. : alk. paper) | ISBN 9780226604398 (e-book)
Subjects: LCSH: Maxwell Street (Chicago, Ill.) | Chicago (Ill.)—Social life and customs.
Classification: LCC F548.67.M39 C74 2019 | DDC 977.3/11—dc23
LC record available at https://lccn.loc.gov/2018049441

♾ This paper meets the requirements of ANSI/NISO Z39.48-1992 (Permanence of Paper).

Contents

Acknowledgments

This book has been over a decade in the making and has morphed through many forms. I started on this project as a member of the geography department at the University of Wales Aberystwyth (now Aberystwyth University) and carried on the work while at Royal Holloway, University of London, and at Northeastern University in Boston. The finishing touches have been completed at Trinity College in Hartford, Connecticut. Needless to say, many people from these institutions and beyond have been helpful along the way. Particular thanks are due to two Maxwell Street enthusiasts who appear in this book—Steve Balkin and Lori Grove—both of whom donated time, information, walks, and energy to this project on more than one occasion. I acknowledge the assistance of the Maxwell Street Foundation (previously the Maxwell Street Historic Preservation Coalition). The special collections librarians in the various archives at the Chicago History Museum, the University of Illinois at Chicago, the University of Chicago, the Harold Washington Library, and Northern Illinois University have all been invaluable. I am enormously grateful to Mary Laur, my editor at the University of Chicago Press, for her enthusiasm for this project with all its quirks and complications. The book is better than it would have been thanks to the close attention of Joel Score, also at the Press.

Elements of the book have appeared in different form in the journals *Transactions of the Institute of British Geographers* and *Annals of the Association of American Geographers*; in the book *Mobilizing Hospitality: The Ethics of Social Relations in a Mobile World*, edited by Jennie Germann Molz; and in volume 1 of the *Sage Handbook of Human Geography*. I am grateful to all referees and editors involved. I have benefited from feedback from audiences at talks given at the Institute of Humanities Research at Arizona State University, the University of Copenhagen, the University of Eichstaett, the Urban Space Research Network Symposium in Heidelberg, the Center for Research in the Arts, Social Sciences and the Humanities at the University of Cambridge, the Street Culture Conference in Turku, the Australian Historical Society annual meeting in Melbourne, the University of Provence in Aix-en-Provence, the University of Amsterdam, the Pratt Institute in New York City, the University of Berne, the Center for Interdisciplinary Studies in Society and Culture at Concordia University, Montreal, and the International Colloquium in GeoHumanities at the Universitat Pompeau Fabra in Barcelona. Research for this book was generously supported by an AHRC research fellowship, which provided valuable time to research and write. One piece of advice for writers that I recall: to write well you have to do everything else badly. I hope this has not been noticed too much by my wonderful, adventurous family, Carol, Owen, Alice, and Madison Jennings. You are constant sources of wonder and inspiration.

FIGURE 1.
Corner of Maxwell and Halsted Streets, Chicago, April 1941. Photo by Russell Lee. LC-USF33-012984-M5. Library of Congress, Prints and Photographs Division, FSA/OWI Collection.

PART ONE

Writing Place

How to write about a place? Where do we start? One way is to plot our location—to specify its coordinates:

> 41°51′53″ N, 87°38′49″ W

Or, shifting from global, latitude-longitude coordinates to the local, urban grid:

> the intersection of South Halsted Street and West Maxwell Street

This is a book about the area around this intersection, located in Chicago, a couple of miles south and west of the downtown Loop. It is about the market that existed there for over a century and what became of it.

> The roadways from curb line to curb line of the following streets: West Maxwell Street from the west line of South Union Avenue to the east line of South Sangamon Street, except the roadway of South Halsted Street; West 14th Street and West 14th Place, from the west line of South Halsted Street to the east line of South Sangamon Street.
> **LEGAL DEFINITION OF MAXWELL STREET MARKET, MUNICIPAL CODE OF CHICAGO**[1]

It is a book about the things, stories and practices that gathered there and were dispersed from there.

> The heart of the ghetto is marked by two great thoroughfares: Halsted Street and Maxwell Street. The former is lined on both sides with imposing emporiums: furniture stores, sausage stores, fur stores, cloak and suit, silk and dry goods, shoe, hat and cap, tobacco, and department stores. On Halsted Street business goes on as it would in the Loop. The stores advertise and have one price. Not so with Maxwell Street. Maxwell Street is as native to the ghetto as

> Halsted Street is now foreign to it. On Maxwell Street there is life; on Halsted Street, decorum. Maxwell Street is the Halsted Street of a generation ago. The proprietors of the substantial establishments on Halsted are the graduates of Maxwell, for the most part. The modern business man on Halsted Street represents the ideal of the sons of the pushcart owners on Maxwell Street.
> **LOUIS WIRTH**[2]

This is a book about the specific area around Maxwell and Halsted and about the idea of place. It is about the ways we might think and write about place (this place) and places in general. It is a work of *local theory* or, perhaps, *topology*.

> But where, Benjamin wonders, does the thought about place take place, or what is the place of place, understood as a philosophical concept? When one reads theoretical musings about the subject, one usually gets the impression that they could have been written anywhere and be about anywhere. Benjamin shows very little interest in ideas *about* place, Instead, he draws his attention to the ideas arising from *this* place. His is one of the rare philosophical works that begins with the question *where* rather than the usual *what* or *how* or *why*. Call it, for the time being, a *topology*, or a theory in situ. Benjamin calls it "presence of mind."
> **DAVID KISHIK**[3]

There is something a little archaic, even quaint, about a geographer writing a book about a place. I have written plenty about "place" as an idea. I have written about specific places in passing. I have never tried to write a book about a particular place. Even critiques of the universalism of metatheory rarely fall back on the particular as a refuge. But here I want to linger a while in all that a place has to offer. I want to offer one version of what place-writing can look like in the early twenty-first century.

Local Theory/ Place-Writing

Local theory is both a way of engaging with place and a way of doing theory. It is different from ways of engaging place that focus on uniqueness and particularity—separating one place from all others. It is also different from constructing theory that universalizes and generalizes. It is different from traditional forms of place-writing in its willingness to take theoretical tangents from the specificity of a particular locale. It is different from much theory in its insistence on staying where we are—staying located.

There is a persistent belief that writing is the end of a process. Read. Research. Write. To me, writing is all of these things at once. Writing is my primary method. The idea of "writing up" research makes little sense to me.

As academics we still carry with us the trappings of science. Some form of detachment is necessary for our work to be properly academic. Writing about a place is supposed to be more than an act of listing its contents. We are called to construct narratives and provide interpretations.

The act of writing is part of the process of relating to place—not just a record of it.

Jane Rendell has invented a process she calls "site-writing." It is a practice of art criticism that acknowledges that the critic and the art are both situated. There is no view-from-nowhere, no god-trick that allows an easy narrative.

> Site-Writing explores the position of the critic, not only in relation to art objects, architectural spaces and theoretical ideas, but also through the site of writing itself, investigating the limits of criticism, and asking what it is possible for a critic to say about an artist, a work, the site of a work and the critic herself and for the writing to still "count" as criticism.
> **JANE RENDELL**[4]

While working on Maxwell Street for over a decade I have been taking notes, copying pieces of texts in notebooks and on my computer (on at least three computers—the project outlasts the hardware). When I write out an interesting paragraph from the archive in a notebook, that is part of the research process. Writing enables me to think.

One way of writing place, prevalent in the first half of the twentieth century, was the regional monograph associated with regional geography and local history. These accounts tended to follow a pattern, a structure that entailed a common understanding of how knowledge was divided. They would start with the physical geography of a place—bedrock, soil types, topography, climate—and move on to human history: first settlers, important figures, incomers. Human geography would follow—house types, land use, boundaries, forms of government, settlement patterns, main crops. Key processes such as agricultural revolutions, urbanization, and industrialization would find their place in the narrative. Toward the end of the book we would encounter the arena of what might broadly be called "culture"—rituals, practices, beliefs.

Description

The narrative of a place that starts with rocks and ends with beliefs involves a number of assumptions. One assumption is that the place has an identity that is not shared with the places around it. It is a unique amalgam of all the items on the list—rocks, weather, soil, plants, buildings, practices, beliefs—a study in particularity. Another assumption is that there is some connection between the bedrock and the beliefs—a continuity through connection between "nature," "economy," and "culture." Similarly, the present of a place can be apprehended through its past.

> To express this another way: the theological motif of calling things by their names tends to switch into the wide-eyed presentation of mere facts.
> **THEODOR ADORNO**[5]

While the tradition of place-writing has never ceased, it became suspect in the social sciences from the late 1950s onward. Such writing rarely said much more than "this place is like this," and this was never enough for science-minded social scientists. Yet much was lost in the setting aside of regional writing within geography and elsewhere. While many regional monographs were not well written, there was, at least, a degree of attention to the act of writing. At their best, they were, indeed, "the highest form of the geographer's art."[6]

> It is a humiliating experience for a geographer to try to describe even a small tract of country in such a way as to convey to the reader a true likeness of the reality. Such description falls so easily into inventory form in which one unrelated fact succeeds another monotonously. How difficult it is to transcend a painstaking compilation of facts by an illuminating image.
> **H. C. DARBY**[7]

> It is not against a body of uninterpreted data, radically thinned descriptions, that we must measure the cogency of our explications, but against the power of the scientific imagination to bring us into touch with the lives of strangers. It is not worth it, as Thoreau said, to go round the world to count the cats in Zanzibar.
> **CLIFFORD GEERTZ**[8]

It is not just "science" that has brought the act of description into disrepute. Even advocates of description have insisted on the need to add a layer of interpretation—to produce "thick" in place of "radically thinned descriptions." Clifford Geertz famously argued for an act of description that would help to build an anthropological version of theory—"because the essential task of theory building here is not to codify abstract regularities but to make thick description possible, not to generalize across cases but to generalize within them."[9]

Perhaps it is the case that a "painstaking compilation of facts" might enact a kind of transcendence of its own. Perhaps description need not bow down to explanation. It was precisely the compilation and juxtaposition of facts and observations in the Paris arcades that Benjamin believed might provoke moments of "illumination."

> I have that continuous uncomfortable feeling of "things" in the head like icebergs or rocks or awkwardly shaped pieces of furniture—it's as if all the nouns were there but the verbs were lacking—if you know what I mean, And I can't help having the theory that if they are joggled around hard enough and long enough some bit of electricity will occur, just by friction, that will arrange everything.
> **ELIZABETH BISHOP**[10]

In *Species of Spaces* Georges Perec asks us (readers, writers) to account for places through the construction of lists. Such list-making, he argues, will be dull and mundane but will eventually provide a spark—a moment where we are teleported to a different place and the extraordinary emerges. In his essay on "the street" he urges us to make lists endlessly "You must set about it more slowly, almost stupidly. Force yourself to write down what is of no interest, what is most obvious, most common, most colourless." His instructions become quite precise:

> The street: try to describe the street, what it's made of, what it's used for. The people in the street. The cars. What sort of cars? The buildings: note that they're on the comfortable, well-heeled side. Distinguish residential from official buildings.
>
> The shops. What do they sell in the shops? There are no food shops, Oh yes, there's a baker's. Ask yourself where the locals do their shopping.
>
> The cafés. How many cafés are there? One, two, three, four. Why did you choose this one? Because you know it, because it's in the sun, because it sells cigarettes. The other shops; antique shops, clothes, hi-fi, etc. Don't say, don't write "etc." Make an effort to exhaust the subject, even if that seems grotesque, or pointless, or stupid. You still haven't looked at anything, you've merely picked out what you've long ago picked out.
>
> Force yourself to see more flatly.
>
> **GEORGES PEREC**[11]

"Carry on" (making lists), he writes, "until the scene becomes improbable, until you have the impression, for the briefest of moments, that you are in a strange town or, better still, until you can no longer understand what is happening or what is not happening, until the whole place becomes strange, and you no longer even know that this is what is called a town, a street, buildings, pavements."[12]

The last decade or so has seen a revival in place-writing across the humanities and social sciences, as well as in creative nonfiction. Recent writing of place (and region) has taken an experimental turn—the complexities of place-as-assemblage have been reflected in various kinds of text-as-assemblage.[13]

> Far from being a dull inventory trap, description offers complex cultural practice, at once calculating and generative, giving an account of landscape in the sense of both patient itemisation and events proceeding. Description carries a usefully dual performative sense of, on the one hand, distanced representation—the observer set back, however closely, from a scene—and on the other inscriptive enactment—the instrument describing, engraving, a line.
>
> **DAVID MATLESS**[14]

The work of assemblage is a kind of description, a constant deferral of explanation.

Writing place here enacts a particular kind of description. I do not often describe Maxwell Street as I have encountered it. Rather I perform a description of descriptions—an account of a place over a hundred years through the descriptions of others. These descriptions are themselves kinds of performances—they act on place as much as they account for it.

An assemblage is a whole made up from parts in which the parts can change while the whole remains. Maxwell Street was, like all places, an assemblage. It was a gathering place of things, meanings, and practices. This gathering quality of place necessitates writing in a way that follows the unruly trajectories that pass through it. Maxwell Street is a starting point for thinking and writing about aesthetics, materiality, narrative, performance, waste, the senses, value, memory, the blues, and tax increment financing. Among other things.

Parataxis

> When objects are put together, we cannot attribute to them coordinations as complicated as in human language. In reality, the objects—whether these are the objects of the image or the real objects of a room, or of a street—are linked only by the single form of connection, which is parataxis, i.e., the pure and simple juxtaposition of elements. This kind of parataxis of objects is extremely frequent in life: it is the system to which are subject, for example, all the pieces of furniture in a room. The furnishing of a room achieves a final meaning (a "style") solely by the juxtaposition of elements.
> **ROLAND BARTHES**[15]

Parataxis refers to the placing of things side by side.

> Clucking white pullets, geese, pigeons, rabbits and pet pups. New straw hats and vintage bird cages. Musical instruments, fresh strawberries, crockery, ladies' hats in the latest cuckoo designs and kerosene lamps.
> ***CHICAGO DAILY NEWS*, 1939**[16]

There is a sense in which places are *paratactic*. While places often have narratives attached to them, they can also be thought of as lists—as one thing and then another. Like description, the paratactic has commonly been thought of as inferior to proper syntax—a language structure that makes causality and logic clear.

> Although it might seem as if writing in the additive style is just a matter of putting one thing after another in no particular order (how can that be hard?), it is in fact the far more difficult style to master; for the relative absence of formal constraints means that there are no rules or recipes for what to do because there are no rules or recipes for what not to do.
> **STANLEY FISH**[17]

Paratactic writing is the creative equivalent of a flat ontology. Components are leveled. Nothing is subordinated.

> So: start in the middle and read outward, start in the middle and read upward; it is yours to make: design, whittle, cut, snip, tie, glue, trim, rasp, paint,

grow vexed, cuss, and pitch it across the room (we will then share one more thing): it is yours to show how the pieces can fit together, perhaps even to demonstrate how the job *should* be done.

WILLIAM LEAST HEAT-MOON[18]

In linguistic terms, once might say that the figures are distributional but not integrative; they always remain on the same level: the lover speaks in bundles of sentences but does not integrate these sentences on a higher level, into a work; his is a horizontal discourse: no transcendence, no deliverance.

ROLAND BARTHES[19]

Nonlinear Writing

Beyond traditional place-writing are other lineages, experiments in writing that suggest nonlinear forms for histories and geographies.

Unlike most scholarly books, what follows is a riot of short chapters. I wanted them to be like the flushes of mushrooms that come up after a rain: an over-the-top bounty: a temptation to explore: an always too many. The chapters build an open-ended assemblage, not a logical machine; they gesture to the so-much-more out there. They tangle with and interrupt each other—mimicking the patchiness of the world I am trying to describe.

ANNA LOWENHAUPT TSING[20]

This book, then, is not a smooth story that follows the lines of its own progress from beginning to end as a master narrative would but a collection of fits and starts in the moves of master narrative itself.

KATHLEEN STEWART[21]

Writing is rarely part of methodology courses in the social sciences. Why not?

Contemporary poets cross the lines between poetry and prose in order to delve into the complexities of place, history, and culture. Claudia Rankine's segmented prose paragraphs provide a powerful account of racist micro-aggressions. Susan Howe mixes poetry and prose with paratextual techniques to explore moments in American history (footnotes as creative strategy!). Maggie Nelson uses discrete, sometimes numbered text blocks in her critical-creative engagements with transsexuality, motherhood, sex, murder, and the color blue.[22] The boundaries between criticism, prose, and poetry have been blurred. Such explorations have informed the present text.

A day or two after my love pronouncement, now feral with vulnerability, I sent you the passage from *Roland Barthes by Roland Barthes* in which Barthes describes how the subject who utters the phrase "I love you" is like "the Argonaut renewing his ship during its voyage without changing its name." Just as the *Argo*'s parts may be replaced over time but the boat is still called the *Argo*, whenever the lover utters the phrase "I love you," its meaning must be renewed by each use, as "the very task of love and of language is to give to one and the same phrase inflections which will be forever new."

MAGGIE NELSON[23]

While not writing exactly, Richard McGuire's graphic text *Here* is one of the most successful explorations of place. All of its images occur in the same location—for much of time, the site of the living room of a grand home. We see geology, biology, indigenous folk in the woods, a newborn baby, ballet. But not in that, or any, order. An image from one moment is inset into an image of another. Space and time collapse. No explanations or interpretations are given—nor are they needed.

> But who or what is the main character? Is it the man who seizes up at a joke told in the first few pages (yet dies, moments later, halfway through the book, after the reader has already ricocheted back and forth through millions of years of history)? Is it the indigenous couple, looking for a place to copulate? Is it the cat, the cat's cradle, the elk, the builders, the partygoers, the weeping woman? You could say it's the space of the room, the arbitrary geometry imposed by a human mind on a space for reasons of shelter and as a background to this theatre of life. But you could also claim it is the reader, your consciousness where everything is pieced together and tries to find, and to understand, itself.
>
> **CHRIS WARE**[24]

Chris Ware's own boxed set of graphic stories, *Building Stories*, explores episodes whose main connection is that they take place in a single building. The accumulation of strivings and (mostly) failures is unbearable.[25] They do not form a narrative—they enact spatial coincidence.

Montage

This book takes inspiration both from Walter Benjamin's use of montage in *The Arcades Project* and from the idea of the "commonplace book," also used by Bruce Chatwin in *The Songlines* and by William Least Heat-Moon in *PrairyErth*.[26]

Montage is a close cousin of parataxis. It refers to the juxtaposition of disparate elements—a kind of assembling. Geographer Allan Pred used montage to interrupt the business-as-usual construction of metanarratives in theory. He did this in an experimental mode, breaking up his text such that it had the appearance of poetry. There was, however, little logic to the lineation. His text was prose and could just as easily have been written as conventional paragraphs. Nevertheless, there is some of the spirit of Pred in what follows.

> Through assembling (choice) bits
>   and (otherwise neglected of discarded) scraps,
> through the cut-and-paste reconstruction of montage,
> one may bring alive,

> open the text to multiple ways of knowing
> and multiple sets of meaning,
> allow multiple voices to be heard,
> to speak to (or past) each other
> as well as to the contexts from which they emerge
> and to which they contribute.
>
> **ALLAN PRED**[27]

> Method of this project: literary montage. I needn't *say* anything. Merely show. I shall purloin no valuables, appropriate no ingenious formulations. But the rags, the refuse—these I will not inventory but allow, in the only way possible, to come into their own: by making use of them.
>
> **WALTER BENJAMIN**[28]

The Arcades Project is a hefty bundle of fragments and files. Benjamin combines his own pithy ruminations with snippets and notes from the writings of others. His own thoughts rarely refer directly to his found texts. The moments of illumination sparked by his juxtapositions of fragments are not the satisfying revelations of a story well told but flashes of recognition that come from the side by side and misaligned.

Conceptual poet Kenneth Goldsmith attempts to outdo Walter Benjamin. In *Capital*, he collects snippets and fragments about New York City and presents them without commentary or synthesis. We can read Goldsmith through his choice of snippets and chapter, or *convolute*, headings.[29]

> This is an exercise that is as much about reading as it is about writing. In fact, it's a book that proposes reading as writing. While many other authors could have written Benjamin's book, why is it that his is so successful, so endlessly fascinating? It's about what he chooses. In lesser hands, such a work would've been dreary, dull, tedious, pedantic and loquacious. Instead, we have a book that is arguably one of the most readable works ever written, yet very few words were actually written by Benjamin himself; it's an act of conceptual writing where what one chooses—one's taste—either makes or breaks the book. While there was much Benjaminian gloss and "voice" in his book, I take it to the next level: nowhere is a single word of my own present—not a thought, not a commentary, nor a sentiment—instead, reflecting contemporary concerns, my task is merely appropriative.
>
> **KENNETH GOLDSMITH**[30]

Juxtaposition is different from randomness. Benjamin juxtaposes with purpose. Connections are implied by spatial contiguity.

There is a pleasing symmetry between the montage form and how places present themselves: one thing after another, their relations implied by spatial arrangement. Montage sits in an uneasy relation to the linearity of narrative.

Like all historians, we configure the events of the past into causal sequences—stories—that order and simplify those events to give them new meanings. We do so because narrative is the chief literary form that tries to find meaning in an overwhelmingly crowded and disordered chronological reality. When we choose a plot to order our environmental histories, we give them a unity that neither nature nor the past possesses so clearly. In so doing, we move well beyond nature into the intensely human realm of value.
WILLIAM CRONON[31]

But it is precisely because the narrative mode of representation is so natural to human consciousness, so much an aspect of everyday speech and ordinary discourse, that its use in any field of study aspiring to the status of a science must be suspect. For whatever else a science may be, it is also a practice which must be as critical about the way it describes its objects of study as it is about the way it explains their structures and processes.
HAYDEN WHITE[32]

Narrative demands a beginning, a middle, and an end. Such an arc gives shape to the carnival of facts and things that make up a place or landscape. Narratives please us with their seemingly effortless ordering and moral trajectories.

While not strictly montage, this book is informed by montage. Themes emerge, submerge, and are repeated. The technique is sedimentary and rhythmic. While it is not a linear narrative it is certainly possible to construct narratives from its contents.

Commonplacing

Time was when readers kept commonplace books. Whenever they came across a pithy passage, they copied it into a notebook under an appropriate heading, adding observations made in the course of daily life. Erasmus instructed them how to do it; and if they did not have access to his popular *De Copia*, they consulted printed models or the local schoolmaster. The practice spread everywhere in early modern England, among ordinary readers as well as famous writers like Francis Bacon, Ben Jonson, John Milton, and John Locke. It involved a special way of taking in the printed word. Unlike modern readers, who follow the flow of a narrative from beginning to end, early modern Englishmen read in fits and starts and jumped from book to book. They broke texts into fragments and assembled them into new patterns by transcribing them in different sections of their notebooks. Then they reread the copies and rearranged the patterns while adding more excerpts. Reading and writing were therefore inseparable activities. They belonged to a continuous effort to make sense of things, for the world was full of signs: you could read your way through it; and by keeping an account of your readings, you made a book of your own, one stamped with your personality.
ROBERT DARNTON[33]

Appropriately enough, compiling a commonplace book is called "commonplacing"—producing a literary *topos*. Commonplacing is a way of recording the wisdom of others, and gradually drawing upon it to form a unique new assemblage—your own kind of illuminating wisdom. In many ways, Benjamin's *Arcades Project*, with its moments of illumination, is an ideal modern type of the practice.

I suspect that many writers in the humanities engage in commonplacing, even if unintentionally. We read and write as method, collecting fragments in archives. As we take a walk, or drive to work, the fragments shift and achieve momentary form—form that is often lost by the time we get to our destination.

Topopoetics

Every now and then, geographers nervously intimate that their discipline might aspire to the status of art. "Creative geographies" are emerging.[34]

> Geography will deserve to be called an art only when a substantial number of geographers become artists.
> **DONALD MEINIG**[35]

The word *geography* means "earth writing," yet relatively little attention has been paid to the "writing" part.[36] I am interested in writing as making.

The strategy used in this book might be called *topopoetics*.[37] *Topo* comes from *topos* (τόπος), the Greek for "place." This is combined with *poetics*, which comes from *poiesis* (ποίησις), the ancient Greek term for "making." *Topopoetics* is thus "place-making."

In the philosophy of Aristotle, *topos* appears both in accounts of how the world comes into being and as a figure in rhetoric. In rhetoric, a *topos* is a "particular argumentative form or pattern."[38] Like a form in poetry—a sonnet or a villanelle—it has a particular shape. This rhetorical view of *topos* is linked to the world through the art of memorizing long lists by locating the listed items in particular *places*. In Aristotle's rhetoric, it is important to choose the right kind of *topos* for the argument at hand, just as it is important to select the right form for a poem.

> For just as in the art of remembering, the mere mention of the places instantly makes us recall the things, so these will make us more apt at deductions through looking to these defined premises in order of enumeration.
> **ARISTOTLE**[39]

Topos also points toward the centrality of place for being. To be is to be *in place*—to be *here* or *there*. The two meanings of *topos*—as rhetorical form and as place—combine in the project of topopoetics. Writing is a kind of building. We are making place. Part of that process is finding the correct form to convey a place.

> The "place" or *topos* of a thing is thus understood to be the inner surface of the body (where "body" here means simply the thing in its physical extendedness) within which that thing is enclosed—on this account the "place" of a rosebud contained within a glass paperweight is the inner surface of the glass that surrounds the enclosed flower. The implication of this account is that to be "in place" is always to be contained with an enclosing body.
> **JEFF MALPAS**[40]

> The opposite of parataxis is hypotaxis, the marking of relations between propositions and clause by connectives that point backward or forward. One kind of prose is additive—here's this and now here's that; the other asks the reader or hearer to hold in suspension the components of an argument that will not fully emerge until the final word. It is the difference between walking through a museum and stopping as long as you like at each picture, and being hurried along by a guide who wants you to see what you're looking at as a stage in a developmental arc she is eager to trace for you.
> **STANLEY FISH**[41]

Imagine for a moment a blank page with just the words "Maxwell Street" written on it. Better yet, take a piece of paper and write "Maxwell Street" on it. Stare at it a while. Consider the difference between the blank sheet and the sheet you now have in front of you. Topopoetics is happening.

My aim has been to write this book in a way that reflects both the experience of place—and particularly Maxwell Street—and the experience of researching place. My hope is that the structure of the book encourages a different type of reading.

This is not a linear book. The structure of the text is intended to reflect the capacity of place to gather and disperse stories and narratives as well as things and practices. Fragments of ideas flow through place and punctuate this text. The text is divided into paragraphs (and images) with spaces between them. The paragraphs accumulate into an argument through description and addition. Many of these paragraphs are direct quotations on the model of the commonplace book.

> Because of the deliberate unconnectedness of these constructions, Benjamin's insights are not—and never would have been—lodged in a rigid narrational or discursive structure. Instead, they are easily moved about in changing arrangements and trial combinations, in response to the altered demands of the changing "present."
> **SUSAN BUCK-MORSS**[42]

A second kind of quotation is what we might call evidence—nuggets of information gleaned from archival research into the hundred-year history of a particular place—the Maxwell Street Market in Chicago.

The roadways from curb line to curb line of the following streets . . .

In so far as this exercise is haunted by the spirit of Walter Benjamin, the market serves as my Paris arcades. These nuggets were collected over a decade of visits to Chicago and elsewhere.

> I am not invoking guarantees, merely recalling, by a kind of salute given in passing, what has seduced, convinced, or what has momentarily given the delight of understanding (of being understood?). Therefore, these reminders of reading, of listening, have been left in the frequently uncertain, incompleted state suitable to a discourse whose occasion is indeed the memory of the sites (books, encounters) where such and such a thing has been read, spoken, heard. For if the author here lends his "culture" to the amorous subject, in exchange the amorous subject affords him the innocence of his image-repertoire, indifferent to the proprieties of knowledge.
> **ROLAND BARTHES**[43]

The book has three parts. The first is this reflection on place-writing. The second is an account of Maxwell Street—an act of writing place. The third and final section is an outline of a theory of place derived from the account of Maxwell Street.

Putting "theory" at the end allows us to enter the theorization of place informed by our wanderings through Maxwell Street—reversing how things are normally done. Place gathers and disperses things, stories, and practices but also ideas. Local theory involves connecting what is there—in Maxwell Street—to what lies beyond. It traces unruly trajectories in and out of place.

The writing strategy applied here attempts to present an account of place that is gently nonlinear—a spiral, perhaps, moving in and around, engaging the same location from different perspectives. Such a strategy, I believe, can allow academics to access the creative mind, and readers to get productively lost.

> To lose yourself: a voluptuous surrender, lost in your arms, lost to the world, utterly immersed in what is present so that its surroundings fade away. In Benjamin's terms, to be lost is to be fully present, and to be fully present is to be capable of being in uncertainty and mystery. And one does not get lost but loses oneself, with the implication that it is a conscious choice, a chosen surrender, a psychic state achievable through geography.
> **REBECCA SOLNIT**[44]

On the edge of the black, Greek, and Italian neighborhoods, we discover, by chance, one of the biggest markets I've seen since the Djema el Fna Square in Marrakech. It stretches for more than a mile along the pavement and sidewalks of a broad street. The sun is punishing, a midsummer sun, and dark Chicago has suddenly become an exotic village, hot and colorful, the way I'd imagined San Francisco would be.

SIMONE DE BEAUVOIR[45]

My approach is interdisciplinary. No carefully formed hypothesis generated the contents of this book. Knowledge here is emergent. The writing is method. It did not happen at the end of a process—it was the beginning, middle, and end.

> Culture was reconceived as an assemblage of disparate and incommensurate things throwing themselves together in scenes, acts, encounters, performances, and situations. Writing became an attunement, a response, a vigilant protection of a worlding. Both writing and culture became potentially generative and capacious. A writing might skid over the surface of something throwing itself together or it might pause on a strand as it moved with other strands or fell out of sync, becoming an anomaly or a problem. Writing could be a way of thinking.
>
> **KATHLEEN STEWART**[46]

Despite the particularity of this endeavor I retain the hope that this book may inform more general explorations of place. I have long held to the paradox of the universal and the particular. The universal is that which is shared by everyone—the objective, the omnipotent—and as such is experienced by no one. The particular, by contrast, is subjective and humble, specific to a situated self—yet, in some form, is experienced by everyone. It is for this reason that reading a book about a place we have never been to can inform the way we look at our own place. We recognize the universality of the experience of the particular—the experience of place.

Maxwell Street is a special kind of place. It is highly represented and thus possible to explore in and through an extensive archive. It has its novelists, poets, and philosophers. Journalists, planners, and politicians visited. Activists accumulated documents and things in an effort to save it. Photographs of the market fill thick files in the Chicago History Museum. Not all places are so archivally present. It is not clear that a book on a place ten blocks to the west or south could be written in the same way. Nevertheless, Maxwell Street is hardly globally famous. It is not the Empire State Building or Trafalgar Square. I doubt many of my friends in London or even Boston would have ever encountered Maxwell Street other than through my growing enthusiasm. I contend that any place has the basic attributes of Maxwell Street—a material presence, stories and narratives, and specific practices. The method of exploration may change but the elements are always in place. All places can be written.

There are three operations I typically engage in during the process of (academic) writing. I gather "evidence" in the form of what might be called primary or archival material. I read around a subject and collect passages that speak to me and to the topic at hand. And then I write passages that link the archival nuggets and commonplace pickings. I try and construct a compelling story—an argument that links pieces of evidence in a more-or-less convincing way. I am not a scientist. I have no illusions about my work being "representative," "replicable," or "falsifiable."

This book has been a ten-year writing exercise conducted in museums, libraries, offices, homes, and hotels, on airplanes and in airports, in the Caribou coffee shop on the corner of Maxwell and Halsted. This book was formed in particular places.

> As I collected blues for this project—in folders, in boxes, in notebooks, in memory—I imagined creating a blue tome, an encyclopedia of blue observations, thoughts, and facts. But as I lay out my collection now, what strikes me most is its *anemia*—an anemia that seems to stand in direct proportion to my zeal. I thought I had collected enough blue to build a mountain, albeit one of detritus. But it seems to me now as if I have stumbled upon a pile of thin blue gels scattered on the stage long after the show has come and gone; the set, striked.
>
> **MAGGIE NELSON**[47]

When I started this project, I had no idea that I would pursue a PhD in creative writing and become a published poet—that I would take a course with the Faber Academy or attend the writing studio at the Banff Center for Arts and Creativity. When I passed through immigration at Calgary International Airport, the immigration officer asked my reason for visiting. I said I was a "writer." I had no idea that I would buy a book called *Bluets* in a bookshop in Seattle.

The work of writing typically seeks to make the connections between evidence, secondary texts, and authorial voice as invisible as possible. The "good" writer makes these links appear as natural as possible. My aim here is to make these workings clear—to make the infrastructure visible.

> A new sentence is more or less ordinary itself but gains its effect by being placed next to another sentence to which it has tangential relevance. New sentences are not subordinated to a larger narrative frame nor are they thrown together at random. Parataxis is crucial: the internal, autonomous meaning of a new sentence is heightened, questioned, and changed by the degree of separation or connection that the reader perceives with regard to the surrounding sentences.
>
> **BOB PERELMAN**[48]

Beginnings

How to begin writing a place? How might an account of the area around Maxwell Street start?

> Before the city, there was the land. Go back just over a century and a half to the place that became Chicago, and our familiar distinction between city and country vanishes. At the mouth of the river where the city would one day stand, small human settlements came and went, but their inhabitants would no more have used the word "urban" to describe the place then the word "rural." Without those words, there could be no city here, not until people came who could dream city dreams in the midst of a cityless landscape. Chicago remained a gathering place like so many other gathering places scattered between the Great Lakes and the Rocky Mountains. What most distinguished it were the wild garlic plants that grew amid the grasses and sedges of its low-lying prairie. From them it gained its name: Chigagou, "the wild-garlic place."
>
> **WILLIAM CRONON**[49]

Rather than providing an account of a place with clear boundaries developing through time, Cronon follows connections with elsewhere—the paths in and out of town.[50] Nevertheless, he narrates the necessary deep history that provided the possibility of Chicago becoming a gathering place. We hear of the glaciers that, at the end of the last ice age, formed a terminal moraine, a ridge that divided two watersheds—the water on one side flowing to the Mississippi River, and on the other into the Great Lakes Basin. What was not yet Chicago was already a meeting place. The Algonquian Indians gave this place a name, attributing importance to a muddy river where wild garlic grew. They may have called the river "Shikaakwa"—their name for the plant—and this may have been translated into "Checagou" by the explorer Robert de la Salle in 1687.

By 1830 Chicago was still a very small trading post where local tribes traded with each other and with the French and English. Around two hundred people called Chicago home. One of these people was Dr. Philip Maxwell.

Following the Blackhawk War of 1832 and the arrival of soldiers from the east, reports of the fertility of the land around Chicago reached the east coast. Over the next several decades Chicago would grow into the second largest city in the United States. By 1880 around half a million people lived there—and an informal market of Jewish street peddlers had emerged at the intersection of West Maxwell Street and South Halsted Street. At 41°51'53" N, 87°38'49" W.

Maxwell Street is on Algonquian land.

Dr. Philip Maxwell, a former member of the New York state assembly, acted as a physician in Fort Dearborn in the 1830s. He signed the Chicago Treaty as a witness in 1833 and is listed as one of five hundred residents on a census before Chicago was incorporated in 1837. Between 1844 and 1847 he was a physician in the new city, known for his jolly disposition and his rotund stature (weighing in at 280 pounds). In 1853 he became the state treasurer of Illinois.

He was a man of general character, and was ever a welcome guest at all public and social gatherings. He had a commanding form and a dignified bearing, which always insured attention and respect. His name is intimately and honorably connected with the early history of our city.
OBITUARY, *CHICAGO PRESS & TRIBUNE*, 1859[51]

One way the doctor's name remained connected with the rapidly expanding city was in the naming of a street. Maxwell Street first appeared on a map of Chicago in 1847, when the population of the city was around thirty thousand. The original street consisted of wooden boards over mud, flanked by small wooden houses built for Irish workers who were building the Galena and Chicago Union Railroad, which arrived a year after the street. This was a place made from mobility.

How else to begin our account? We might list the names this place and the area around it have born over the years. And again define its boundaries.

Maxwell Street
West Thirteenth Place
the Near West Side
Ellis Island of the Midwest
Jewtown
Roosevelt-Halsted
Roosevelt-Union Redevelopment Area
University Village
University Village Maxwell Street
the Ghetto

West of the Chicago River, in the shadow of the Loop, lies a densely populated rectangle of three- and four-story buildings, containing the greater part of Chicago's immigrant colonies, among them the area called "the ghetto." This area, two miles wide and three miles long, is hemmed in on all sides by acres of railroad tracks. A wide fringe of factories, warehouses, and commercial establishments of all sorts encloses it. It is the most densely populated district of Chicago and contains what is probably the most varied assortment of people to be found in any similar area of the world.
LOUIS WIRTH[52]

We might gather stories of arrival.

The city announced itself to our noses before we ever saw it, and we always pressed our faces against the windows to locate the sweet pungent odor that was Gary. (Gary and Chicago blend in my child's eye view as a single place, united in a child's mythic name: The City.) The forest of smokestacks, the great plumes of white and unwhite steam, were unlike any place that I, middle-class child of a nurse and a professor, had ever lived. The place remains in my memory as a gray landscape with little vegetation, a clouded sky

hovering over dark buildings, and an atmosphere that suddenly made breathing a conscious act. I remember especially one smokestack with dense rusty orange vapor rising like a solid column far into the sky before it dissipated. We always saw it there, every year, and it signaled our entrance into The City.
WILLIAM CRONON[53]

I first encountered Maxwell Street in Madison, Wisconsin. While studying for my doctorate I spent a lot of time on State Street, drinking coffee in Steep & Brew. On the third weekend in July 1987, I found the street jammed with stalls and shoppers. The shops that lined the streets had spilled out and were selling their wares at a discount. This was Maxwell Street Days, the annual "summer sidewalk sale." I had no idea what "Maxwell Street" referred to. I bought a shirt.

Years later I read James Gilbert's book *Perfect Cities: Chicago's Utopias of 1893*.[54] In that book Gilbert considers three attempts to produce ideal forms of order in Chicago's urban space. Being familiar with Chicago, and drawn to the utility and aesthetics of disorder, I wanted to write a counter to Gilbert's book: *Imperfect Cities*. I looked for promiscuous spaces—places where mixing happened daily. Places of impurity. A number of spaces suggested themselves, including Washington Square, across from the Newberry Library, where for decades a wild array of flat-taxers, free-love advocates, and anarchists assembled to argue and pontificate. I was attracted to the nearby Dil Pickle Club—a bohemian club that operated between 1917 and 1935. This club was a home for Wobblies and anarchists as well as less political bohemians who would read poetry and perform plays.

The place that eventually captured my attention was Maxwell Street Market.

I first put two and two together and linked Madison's sidewalk sale with the market in Chicago when my father-in-law told me about being taken to Maxwell Street as a child to have a suit made by one of the Jewish tailors who set up shop along the street. He mentioned the extraordinary experience of members of different races mixing relatively freely, as well as the plethora of foods.

All creeds, all races and just one color: green. The color of money. It was key to the market's long success. For the market stood between neighborhoods, on the Near West Side, in that part of the city that was a transient place and port of entry. And being between places, no one ever claimed it. Here blacks and whites mingled freely, spoke, became friends—long before civil rights—in days when the Loop was all white and the Near South Side only black, here Jewish merchants met the bluesmen. They strung electric cords from the stores to the street to play guitar for the market crowds, and the city air filled with music of the Delta.
ALAN P. MAMOSER[55]

This place was no longer there. I am writing about a place I have never been to.

> If the story ends in a wheatfield that is the happy conclusion of a struggle to transform the landscape, then the most basic requirement of the story is that the earlier form of that landscape must either be neutral or negative in value. It must *deserve* to be transformed.
> **WILLIAM CRONON**[56]

In 2005 I took the elevated train from downtown Chicago to the UIC-Halsted stop. As I exited the train I was surrounded by the roar of traffic from the Interstate 290, which surrounds the station. I climbed the stairs from the platform, passed through the barriers at the top, and emerged on South Halsted Street. From there I headed south a handful of blocks to the intersection of South Halsted and West Maxwell—the epicenter of this account and a place to which we will return many times. On the way, I passed a staggering fortress, the University of Illinois at Chicago campus, home to twenty-five thousand students—all concrete, narrow slit windows, and mock-castle features. It dwarfs the small, redbrick building that is Hull-House, once the home base of the social activist Jane Addams. The building narrowly avoided destruction when the university moved inland from Navy Pier in 1965 as part of Mayor Richard J. Daley's program of "urban renewal." It is now a museum.

A few minutes later I arrived, for the first of many times, at the corner of Maxwell and Halsted.

PART TWO

Market/Place

Maxwell and Halsted

At the corner of Maxwell and Halsted there is nothing much to see. On the many days I've spent time there, the place has been open, sparsely populated, clean and tidy. Maxwell Street itself runs precisely one block on either side of South Halsted Street. On the southwest corner of Maxwell and Halsted is a Caribou coffee shop, part of a Minnesota-based chain. Look west and you'll see the playing fields of the University of Illinois at Chicago. Look east and you'll see a large concrete arch set in a red brick wall. Engraved in the concrete are the words "Maxwell Street." This declaration seems too grand by far for these two blocks of nothing in particular. Above it are two further names—"University Village" and "Marketplace"—as if the area no longer knew what it was called.

Along the south side of Maxwell Street between Halsted and the arch is a row of tidy but mismatched buildings that do not look like those on the blocks that surround them. They seem to shout "History!" Up and down Halsted and along the streets parallel to Maxwell Street are low-rise apartments and town houses, almost all of them recently built. Collectively they have been called University Village by their developers. On my first visit I was met by multiethnic faces gazing benignly from billboards. The billboards disguised the building site for "Ivy Hall" and displayed as well the (high) prices of its units. A 2001 brochure advertised townhomes of between 2,154 and 2,508 square feet $396,000 to $497,000.[1] In 2016 a similarly sized three-bedroom townhome fetches $650,000.

Should you linger you would notice a series of bronze statues and plaques telling something of the history of the place. From the plaques you would learn that this was the site of North America's largest open-air market. You would learn that in the early twentieth century it was one of the most densely populated areas in the United States, a place where immigrants from Europe and migrants from the east and south of the nation settled, a place of cosmopolitan poverty. The bronze statues depict a blues musician and a market

FIGURE 2.
University Village/ MarketPlace/Maxwell Street. Photo by author.

stall seller. There are bronze replicas of packing crates stacked up in mock disorder.

> This is Maxwell and Halsted. The neighborhood awakens. People come out on the street. A parade of hunched shoulders, hard faces, battered felts, baggy pants, ragged coats. Through the hush of half-light they find their way, specters in the city fog of smoke and grit, of lifting dark and spreading dawn. . . . Maxwell Street is a small hub around which a little world revolves. This is Jerusalem. The journey to Africa is only one block. It is one block from Africa to Mexico, from Mexico to Italy two blocks, from Italy to Greece three blocks . . . this is Halsted Street. This is the World street, the Mother street. Mother Halsted is wise. Is patient. She knows their taste, their traditions, their beliefs. She puts the immigrant to sleep in his first New World bed. She holds him in her slum arms. Chicago's most humane street, she adopts them all.
> **WILLARD MOTLEY**[2]

Immigration/ Migration

Some have referred to Maxwell Street area as the "Ellis Island of the Midwest," since for many immigrants to the city it was the first port of call.[3] In the 1920s, when over a third of Chicago's population was foreign-born, the area around the intersection of Maxwell and Halsted Streets was home to immigrants from at least a dozen nations. First settled by Irish railroad workers, the area soon became home to German and Dutch populations fleeing persecution in Europe. By the 1880s it was home as well to a large Jewish population,

FIGURE 3.
Statue of blues musician on Maxwell Street. Photo by author.

FIGURE 4.
Bronze replica market crates with image of blues musician on side. Photo by author.

who quickly established an informal market consisting of peddlers' carts and open stalls selling everything from live chickens to tailored clothes. In 1912 the market was officially recognized as "Maxwell Street Market" by the city council.[4] Four years later the poet Carl Sandburg was capturing a portrait of a vendor selling fish.

> I know a Jew fish crier down on Maxwell Street with a voice like a north wind blowing over corn stubble in January.
> He dangles herring before prospective customers evincing a joy identical with that of Pavlowa dancing.
> His face is that of a man terribly glad to be selling fish, terribly glad that God made fish, and customers to whom he may call his wares from a pushcart.
> **CARL SANDBURG, "FISH CRIER" (1916)[5]**

During World War I Maxwell Street experienced the arrival of African Americans, many of who migrated to Chicago from the American South. Over the next decades, and certainly by the end of World War II, the area became an important center of black culture in the United States. It is regarded as the birthplace of the Chicago blues, a seminal form of urban electric blues, as musicians plugged electric guitars into outlets in the market stalls. By the 1980s the area had become strongly associated with more recent migrants from Mexico and Puerto Rico. Indeed, the current "Maxwell Street Market"—no longer on Maxwell Street—is notable for its excellent Mexican food.

In addition to the Chicago blues, Maxwell Street is credited with contributions to food and fashion. The Maxwell Street Polish—a fried Polish sausage (like kielbasa) in a bun, topped with grilled onions, mustard, and pickled whole peppers—was created by Jimmy Stefanovic, a Macedonian immigrant who worked at a hot dog stand on the corner of Maxwell and Halsted (41°51'53" N, 87°38'49" W) that later became Jim's Original. The zoot suit—initially associated with African American urban communities and popularized by jazz musicians—may or may not have originated in Smoky Joe's, a men's clothier on Maxwell Street. It consists of a long jacket with wide lapels and padded shoulders worn with high-waisted, pegged trousers.

In 1957 the construction of the Dan Ryan Expressway cut Maxwell Street into two, moving the entirety of the market west of Union Avenue. (Jim's Original can now be found hidden away next to an on-ramp.) The expressway, an urban section of Interstates 90 and 94, purposefully divided what was known as the "black belt" (an area of the South Side associated with African Americans) from more prosperous areas inhabited by white people, including the first Mayor Daley. As was often the case in the era of "urban renewal," the construction depended largely on the demolition of black neighborhoods.

In 2008 I visited Gethsemane Missionary Baptist Church with Lori Grove, a local preservationist and Maxwell Street enthusiast. The history of the church reflects that of the area. Built in 1869 as a school for German immigrants, it became a synagogue in 1902. With the arrival of African Americans from the South and elsewhere in the 1930s, and the construction of a new façade, it was transformed into a Baptist church. On the day I visited, there was a service going on, a small but vigorous group of congregants swaying and singing inside. An elderly woman spoke to us on the steps, recalling the days of her childhood, when this had been a center of black life and she had played with other kids in an area now dominated by the Dan Ryan Expressway. The church was demolished in 2015 despite being one of only a handful of extant buildings that had survived the Great Fire of 1871.[6]

The Maxwell Street Market, in its heyday, was the largest outdoor market in the United States. It persisted in one form or another around the intersection of Maxwell and Halsted until 1994, when the nearby University of Illinois at Chicago (UIC) decided to build new playing fields, parking lots, and upmarket apartment buildings on the site. The market was moved in 1994 to nearby Canal Street. In 2008 it was moved again to Desplaines Street. It is still called the Maxwell Street Market.

Writing Maxwell Street

I am not the first person to write about Maxwell Street.

By the 1890s journalists and social workers from Hull-House were visiting the area as part of a general interest in the "ghetto"—at that time a designation associated with the Jewish population.

> Among the dwelling-houses of the Ghetto are found the three types which curse the Chicago workingman,—the small, low, one or two story "pioneer" wooden shanty, erected probably before the street was graded, and hence several feet below the street level; the brick tenement of three or four stories, with insufficient light, bad drainage, no bath, built to obtain the highest possible rent for the smallest possible cubic space; and the third type, the deadly rear tenement, with no light in front, and with the frightful odors of the dirty alley in the rear, too often the workshop of the "sweater," as well as the home of an excessive population. On the narrow pavement of the narrow street in front is found the omnipresent garbage-box, with full measure, pressed down and running over. In all but the severest weather the streets swarm with children day and night.
> **CHARLES ZEUBLIN**[7]

> This story is about little Louis Epstine, aged nine years. As might be guessed, Louie was a Jew. But there are different kinds of Jews. There are Jews who live on Grand Boulevard, and Jews who live on Maxwell Street. For the most part, the Jews on Grand Boulevard own wholesale clothing stores, and, for the most part, the Jews on Maxwell Street work in the stores.
> **CLARENCE DARROW**[8]

> Here were for sale prunes, raisins, nuts, beans, rice, salted herring, soda water, candles, matches, soap, and various other articles too numerous to

mention. All these were uncovered except for the flies. Is it a wonder that this district has more typhoid than any other section of the city?

Nearby was a bench covered with old coats, vests, shoes, and trousers. Here second-hand shoes are bought without a thought as to who the previous wearer was. A woman will pay fifty cents for a pair of shoes and sixty cents for repairing them, whereas for one dollar her boy can be provided with a new pair that has never been worn and carried no possible taint of a contagious disease.
HILDA POLACHECK[9]

Chicago sociologist Charles Zeublin wrote his account of the Maxwell Street "ghetto" in 1895. Lawyer Clarence Darrow and Hull-House worker Hilda Polacheck wrote about the street in the following decade. These were among the first of a long list of writers who have sought to capture something of the place.

Simone de Beauvoir

In 1947 Simone de Beauvoir visited Maxwell Street Market. She was involved in a relationship with the Chicago writer Nelson Algren, and he was keen to show her the city he knew—for the most part a dark underbelly of prostitutes and junkies. Maxwell Street was a stop on his "must-see" itinerary.

The men are wearing shirts of pale blue, delicate green, shrimp pink, salmon, mauve, sulphur yellow, and indigo, and they let their shirts hang outside their trousers. Many women have knotted the ends of their white blouses above their navels, revealing a broad strip of midriff between skirt and bodice. There are many black faces, others olive, tan, and white, often shaded by large straw hats. To the right, to the left, people in wooden stalls sell silk dressing gowns, shoes, cotton dresses, jewelry, blankets, little tables, lemons, hot dogs, scrap iron, furs—an astonishing mixture of junk and solid merchandise, a yard sale mixed with low-price luxury. On the pavement small cars drive around selling ice cream, Coca-Cola, and popcorn; in a glass cage the flickering flame of a little lamp heats the kernels and makes them pop. There's a tiny man in rags with the face of an Indian, wearing a big straw hat and with a hearing device fixed behind his dirty ear; he's telling fortunes with the help of a machine. The apparatus is very complicated, truly magical: on a moving cart there's a glass column full of liquid in which little dolls jump up and down. A customer approaches—a black woman with a naked midriff, who seems at once provocative, skeptical, and intimidated. She puts her hand against the glass; the bottle-imps jump up and sink into the invisible depths of the instrument, which spits out a strip of pink paper on which the customer's fate is written. This mechanical apparatus in the service of magic, the hearing aid beneath the exotic hat, these juxtapositions give this market its unexpected character. It's an eighteenth-century fair in which drugstore products are sold amid the clamor of four or five radios. There's another charlatan worthy of the old Pont Neuf: he has twined a snake around his neck and is selling a black elixir that is supposed to be a cure-all, and he describes the fabulous properties of this universal panacea through a microphone. . . . There are shops on the sidewalks, often below street level, as in the Jewish neighborhood in New York, and the

> merchandise spills out onto the pavement. Through an open door, I see a gypsy covered with scarves and veils in a darkened room; she's kneeling beside a basin, washing her linen. The radios blare, and each is playing a different tune. . . .
>
> Superstitions, science, religion, food, physical and spiritual remedies, rags, silks, popcorn, guitars, radios—what an extraordinary mix of all the civilizations and races that have existed throughout time and space. In the hands of merchants, preachers, and charlatans, the snares sparkle and the street is full of the chatter of thousands of brightly feathered birds. Yet under the blue sky, the grayness of Chicago persists. At the end of the avenue that crosses the glowing bazaar, the pavement and light are the color of water and dust.
>
> **SIMONE DE BEAUVOIR**[10]

Writers of all vocations have found themselves repeating elements of de Beauvoir's account: observing the mix of people from different ethnic backgrounds, the performance of selling or raising money, the cacophony of music and market activity; comparing Maxwell Street to faraway markets in "exotic" locations (in de Beauvoir's case, Marrakech); using lists to convey the confounding jumble of stuff.

De Beauvoir was among the most prominent, and artful, in a stream of writers that flowed continually through the area over a hundred-year period, from the 1880s onward. Many others, from urban planners to journalists to poets and novelists, also sought to convey something of the kind of place Maxwell Street was. Their efforts took many forms: detective novels, doctoral theses, an oral history, an illustrated local history.[11] While these different sorts of writing have different purposes, outlets, and audiences, there are themes that cut across them. Together they attempt to account for a place that is clearly set apart as worthy of attention. They attempt to report a place to others (the readers) who are elsewhere.

Perhaps the first academic account of Maxwell Street was written by a geographer pursuing a master's degree at Northwestern University in 1954.[12]

Reporting

The Chicago History Museum keeps a file of clippings from newspapers about the area around Maxwell Street, with the earliest example being from 1896. At one level, the collection provides a documentary account, a rich chronology of impressions that allows the curious urban detective to chart the market's visible transformations. At another level, it allows us to trace some of the consistent impacts the market has had on journalists—consistencies that link the 1880s to the 1980s.

Perhaps the obvious question is why journalists were visiting Maxwell Street at all. These were Chicago journalists writing for Chicago newspapers to be read by Chicagoans. The accounts are not the kind of "local color" columns, replete with restaurant recommendations and must-see sites, that we encounter in our current daily papers. Many reports, in the early part of the twentieth century, resembled travel writing—accounts of places far away written for readers who were unlikely ever to go there. Later articles still provided intense descriptions of a place far removed from the normal life of most readers, but they also became diagnostic accounts of a place that seemed to be failing—a place doomed to extinction.

One book, written by journalists, in which Maxwell Street appears, is the muckraking pulp-documentary *Chicago: Confidential!*, which claims to deliver the "low down on the big town," as its authors, Jack Lait and Lee Mortimer, also did in *New York: Confidential!* and *Washington: Confidential!*[13] The book, its front matter tells us, is "all true; every statement has been checked and verified. There are those who did want to see this book published. Only those 'in the know' were aware of the real state of affairs in Chicago, and no one dared write it or publish it until Jack Lait and Lee Mortimer told it all—hitting straight from the shoulder and pulling no punches, Never before has there been so much excitement between the covers of one book." Lait was a notable Chicago journalist who had moved to New York to edit the *New York Mirror*, where Mortimer had worked for him as a columnist. The *Confidential!* books were bestsellers until they were discredited in a series of lawsuits in the late 1950s.[14] The books formed part of a cultural trend replicated in movies that claimed to penetrate the underworld of cities, revealing the many layers of vice and corruption that kept them running. It was the search for the lowdown that brought Lait and Mortimer to 41°51'53" N, 87°38'49" W—the corner of South Halsted and West Maxwell Streets, "one of the most colorful cosmopolitan melting-pots of America."

> Around Halsted and Maxwell Streets swarmed the Chicago ghetto, with many thousands of Jews from Russia and eastern Europe, who spoke no English, who arrived with tickets pinned to their clothes and placards hung around their necks. Street and tenements teemed with bearded patriarchs and their families. Stores sprang up with pullers-in on the walks and everywhere were pushcarts which sold everything from garlands to garlic to women's drawers. The Jewish wave followed closely on the Irish, and with Blue Island Avenue the dividing line, there were wonderful ruckuses that kept the Maxwell Street Station house humming.
>
> **JACK LAIT AND LEE MORTIMER**[15]

Lait and Mortimer's visit to Maxwell Street, as part of a muckraking, noirish investigation of a city gone wrong, is indicative of the kind of role Maxwell Street played in writing across genres. Such accounts paint a colorful portrait of a diverse, even foreign, neighborhood. Maxwell Street stands in for the idea of the melting pot—it becomes a metonym for the creation myths of the United States. But it is not all positive. Paradoxically, Maxwell Street also attracts because it is repellent—because it is the kind of place we (the readers) need to know the truth about.

> Negroes live in hovels without roofs, caved in on the sides, steps missing, tilted like miniature towers of Pisa. As many as a hundred live in a shack meant

> for two families. They sleep in hallways, some even standing leaning against stairs. Whole families exist under staircases. Filth overflows to the walks and weedy lots and everywhere junk is piled. At night, Halsted Street thereabouts is a fantastic riot of smells and colors, a jammed jamboree of Negroes, Mexicans, skull-capped Jews, Filipinos and Levantines. From the reeking cafes come rancid odors of cheap cooking. You can buy anything on the street from a girl, price $5, to a stiletto, price $2.50. Street-hawkers sell guns openly at $20, knives, Spanish fly, contraceptives, and obscene pictures and other crude pornography.
>
> **JACK LAIT AND LEE MORTIMER**[16]

Maxwell Street was a privileged site for the diagnosis of the city's ills. It was portrayed as raw and unsanitized. Writers would describe the area as a slab of "reality"—a reality that readers needed to know about even if they would never visit. In this sense, Maxwell Street formed part of a constellation of marginal spaces in American cities where the relatively affluent could engage in the practice of "slumming."

> From the mid-1880s until the outbreak of the Second World War, an overlapping progression of slumming vogues encouraged affluent white Americans to investigate a variety of socially marginalized urban neighborhoods and the diverse populations that inhabited them.
>
> **CHAD HEAP**[17]

This piece of "reality" attracted writers from the wide array of Chicago city newspapers including the *Chicago Daily News*, *Chicago Sun-Times*, *Chicago Tribune*, and, starting in 1971, the free *Chicago Reader*. Their reports to the Chicagoland readership included many of the features of de Beauvoir's account—the paratactic style, the plethora of lists, the engagement of senses of smell and sound as well as sight, the references to "exotic" places elsewhere, and the fascination with the way the market was performed.

Lists

One of de Beauvoir's lists that stands out is a list of the "things" of the market: "To the right, to the left, people in wooden stalls sell silk dressing gowns, shoes, cotton dresses, jewelry, blankets, little tables, lemons, hot dogs, scrap iron, furs—an astonishing mixture of junk and solid merchandise, a yard sale mixed with low-price luxury." Another list includes more abstract and performative elements jumbled up with materials and things to eat: "Superstitions, science, religion, food, physical and spiritual remedies, rags, silks, popcorn, guitars, radios—what an extraordinary mix of all the civilizations and races that have existed throughout time and space." This act of listing reflects, in a spatial form, the experience of the market as a space of mixture. "Rags, silk, popcorn, guitars, radio"—the words appear next to each other just as the things they refer to do in the market. In this sense de Beauvoir's encounter with the market affirms what has been a dominant form of writing about this place. Writers ranging from urban planners to journalists to novelists repeatedly attempted to account for Maxwell Street through the construction of lists. Their lists convey a sense both of the overwhelming quantity of things and of their unlikely, often disorienting juxtapositions. They portray as futile efforts to order what appears as chaos. Materiality makes its presence felt. It was not just people that moved through the market but vast numbers of "things." Over a hundred-year period observers of Maxwell Street were astounded as much by the range of things they encountered as by the range of people and their practices. Again and again, when faced with the market, observers resorted to a poetry of abundance. Long lists of things became a standard trope for journalists. All the many things that passed through the market, from uncertain origins to the domestic spaces of the cities.

> Shoes, clothing, fish, oranges, kettles, glassware, candy, jewelry, vegetables, crates of live poultry, hats, caps, pretzels, hot-dogs, ice cream cones, beads and beans, hardware and soft drinks, lipsticks and garlic are massed together in glorious ensemble of confusion.
> **CHICAGO DAILY NEWS, 1928**[18]

> Wire-fencing in various size rolls; lighted compasses; a bench drill press; spices in industrial-sized containers; refrigerator/freezers; glazed ceramic tiles in boxes; a telephone-answering machine; long-stemmed glasses and crystal goblets; gloves (ski and regular); automobile wheels and tires; underwear; jeans; jackets and jump suits; battery chargers; a snow blower; notebooks and paper for school; baseball trophies; skates (ice and roller); and comforters.
> **CHICAGO TRIBUNE, 1981**[19]

Lists appeared in newspapers as a result of journalists visiting the market to report on what this place was like. Over time the contents of the lists changed. By the 1970s and 1980s the inventory appears to be more mundane and less removed from our twenty-first-century lives. Lists from the 1920s and 1930s feature such things as ladies' hats in cuckoo designs, vintage bird cages, and crates of live poultry. By the time we reach 1981 we vicariously encounter jeans and battery chargers.

It is notable how artfully some of these lists are arranged. They use evocative juxtaposition and poetic techniques such as alliteration in order to capture the sense of abundance that lists induce. "Lipstick" is, I suspect, deliberately placed next to "garlic"; "hardware" is contrasted with "soft drinks"; "beans" are made to mix with "beads." Such lists are not random—they are artful.

These lists have no beginning or end. They intervene in the heart of things—pile thing upon thing in pleasurable ways that seem to point toward the collapse of things as categories. Such lists appear to impose order while simultaneously gesturing toward the impossibility of order. They are like the Ithaca chapter of *Ulysses*, passages of the Old Testament, early English poetry, the *Iliad*.

We could, if we were so inclined, make a list of lists—a taxonomy of categorization. The lists provided by journalists tend toward the kinds of lists a poet might construct. They appear to have been sorted aesthetically, for affect. Sometimes one thing is placed next to another to make the most of the jarring affect of juxtaposition—again, "lipsticks and garlic." At other times the lists hint at classification systems, with blocks of things that are in some way alike: "Clucking white pullets, geese, pigeons, rabbits and pet pups."

Markets are sites for lists. When we go shopping we take lists with us. Were we to run any but the most meager of stalls we would have an inventory. Selling, shopping, consuming, and list-making are intimately related practices. One kind of list of which we have no record is the shopping lists customers undoubtedly took to the market. These are, of course, mundane lists, frequently written on the back of something else, usually thrown away. Nevertheless, even a list of the mundane can, over time, appear as a poetic and nostalgic representation of life being lived.

> Lists narrate practice and desire. They serve as a fulcrum between consumption and destruction, two processes that are so often seen as oppositional. For the list, this seemingly humble and transitory fragment holds clues about objects and possessions, about love and loss, about meaning and memory. Lists can be both ordering and chaotic, trivial and monumental, transitory and haunting. Most significantly, they reveal the power of the mundane and how seemingly ordinary objects can be freighted with huge and unexpected significance.
> **LOUISE CREWE**[20]

Umberto Eco has suggested that there are two kinds of list that pepper the history of representation. One is the kind that asserts order and presents a sense of completeness. These lists (library catalogs, stock inventories, and the like) announce their ability to account for everything. The other kind of list, by contrast, points toward its inevitable incompleteness and suggests the possibility that it could keep going forever. These are lists that suggest infinity.[21] I have encountered both kinds of lists in my explorations of Maxwell Street. But it is the latter that predominate. These are the kinds of list that humanity has frequently made when confronted with the chaotic.

Lists may be the first kind of writing—a series of nouns mapped onto things in the world. It is certainly the case that much early literature features seemingly endless lists of names or places.

One set of lists that is remarkable for its very unremarkable contents is provided by Georges Perec in his attempt to provide an account of place and an account of writing about place. In *An Attempt at Exhausting a Place in Paris*, Perec observes the Place Saint-Sulpice from a number of café windows and notes, in list form, the things he sees outside of the window.

A 63 [a bus] passes by
Six sewer workers (hard hats and high boots) take rue des Canettes.
Two free taxis at the taxi stand
An 87 passes by
A blind man coming from rue des Canettes passes by in front of the café; he's a young man, with a rather confident way of walking.
An 86 passes by
Two men with pipes and black satchels
A man with a black satchel and no pipe
A woman in a wool jacket, smiling
A 96
Another 96
(high heels: bent ankles)
An apple-green 2CV
A 63
A 70
GEORGES PEREC[22]

Such list-making is one way of approaching place, but it is quite obviously also futile and exhausting. Places gather, assemble, and weave, and lists attempt to account for the results of these actions. Perec's experiment has a sense of failure mixed with melancholy about it. It is an attempt to capture, even exhaust, a place. It is a mundane mirror of the *Iliad*. But it does not come even close to its stated aim of exhausting a place.

An Attempt at Exhausting a Place in Chicago

Corner of Maxwell and Halsted Street, 11:50 a.m.–12:50 p.m., Sunday, October 21, 2012. 41°51'53" N, 87°38'49" W (the one hundredth anniversary of the city council's official recognition of Maxwell Street Market).

I sit in Caribou Coffee, on the southwest corner of Maxwell and Halsted.

Large man in a lumberjack shirt and thin woman wearing shades sit on a black metal bench (fake cast iron) with a small Jack Russell
White Toyota Prius parking in the one remaining spot
Opposite: "Morgans—on Maxwell Street, Bar and Grill"
The other corners of the intersection: Jamba Juice, Sheikh Shoes
"I have a small vanilla latte!"
Cars passing in a steady stream, mostly black, white, or silver
Hipster with goatee and cap, black trousers with white lightning strikes down the side, red jacket
Woman with dog (black)
Black and orange cab—"Stay Hydrated Chicago" sign on top—stops opposite
"I have a medium decaf coffee!"
Man in a Bogart hat enters Morgan's
Cyclists with helmets
Jogger in black with beanie hat

Woman on cell phone
Man staring at the screen of his phone walking slowly, not looking up
Canvas signs on lampposts—"University Village"—picture of woman of uncertain ethnicity (white? Asian? Latina?)
"Your village in the City!"
School bus painted black—"Untouchable Tours"
Street lamps painted black—meant to look like old gas-lamps
Surface of Maxwell Street made from what look like red bricks
Hoodies with various college logos
"I have medium skimmed latte!"
8-Halsted bus, with ad for Burger King—original chicken sandwich, buy one, get one free
Man and woman hand in hand—him with White Sox jacket
Two to five people, on average, visible at most times—but briefly can't see anyone
Thin jogger—gray top, shades, luminous running shoes
Hardly a cloud
Still a steady stream of cars
Window frames painted matte green or terra-cotta—tasteful
Red brick, sandstone—"Maxwell Street" inscription above the second-floor windows
18 bus—16th/Cicero
Lots of sporting attire, shades, phones out
Group of three (dad? granddad?)
18 bus—"are you curious" written on the side
"Small, iced, berry mocha, no whip—enjoy your stay!"
Family, three kids, one being carried
Man in North Face jacket, camera
"Two men and a truck—movers who care"
Kid on dad's shoulders; mum looks at Morgan's menu
Family with teenage boy (14?) in a fancy suit. Church getting out?
Backpacks, baseball caps
U-Haul truck—"still as low as $19.95"
Long yellow truck—"expert driving school, Spanish spoken, student driver"
Rose Paving pickup truck (red) towing tar vat
Cars look new on the whole, mainly economy size
Woman running with pram
Church has finished
More cyclists, fewer dogs
UIC sweatshirt, red on black
Some people in shorts, others in wool hats and coats
Three men at bus stop—two students (?) with short hair, athletic tops, backpacks; one older guy with a hoodie
Girl in purple in a stroller playing with beads
Woman, red streak in hair, gray sweats, sitting on bench with friend holding an iPad

Red pickup truck
Blue station wagon parking badly; well-built driver with Hawaiian shirt and baseball cap
Greyhound bus with blue greyhound on its side
A sudden profusion of red cars
More children now. Church?
Silver Macs outnumber PCs in this café
Kid (4?) in blue sweater playing with fire hydrant
8 bus—Halsted/76th
I guess 30 percent of people are looking at phones
Man in green tracksuit with brown and white dog (medium)
"Two percent, skimmed or soy?"
Large (tall) guy in white basketball vest with large X on front crossing intersection
18 bus—Roosevelt.
Prius leaves
Sunny day, but some of the "gas-lamps" are on
More people in shorts now
Black population seems especially dressed-up—must be church
Police car with blue lights flashing—its passage blocked at intersection
"I have a medium skimmed latte!"
More sports team regalia
El Milagro tortilla truck
"Medium iced Americano—you're welcome!"
18 bus—16th/Cicero.
Traffic backing up at the stop sign
"Large chai latte!"
Shoe shop filling up
"I have a large, extra shot, latte!"
8 bus with American Apparel ad
Fewer people, more cars

One of the most influential lists is the Linnaean classification system for living things. This is a nested hierarchical list—splitting the universe into kingdom (animal, vegetable, etc.), phylum, order, and so on, down to genus and species. This is a list in which everything is made to have a place, even if we are not yet aware of its existence. When a new species is discovered, it is slotted into place, as though the place was already there, awaiting the discovery. This works through forms of likeness. Things that resemble each other in ways that are deemed important are placed close to each other on the list. The periodic table organizes the list of elements in a similar way. But lists can also work aesthetically by placing things next to each other that produces the shocking affect of juxtaposition. Poets often use lists in this way.

This passage quotes a "certain Chinese encyclopedia" in which it is written that "animals are divided into: (a) belonging to the Emperor, (b) embalmed,

> (c) tame, (d) sucking pigs, (e) sirens, (f) fabulous, (g) stray dogs, (h) included in the present classification, (i) frenzied, (j) innumerable, (k) drawn with a very fine camelhair brush, (l) et cetera, (m) having just broken the water pitcher, (n) that from a long way off look like flies." In the wonderment of this taxonomy, the thing we apprehend in one great leap, the thing that, by means of fable, is demonstrated as the exotic charm of another system of thought, is the limitation of our own, the stark impossibility of thinking *that*.
>
> **MICHEL FOUCAULT**[23]

Foucault's consideration of Jose Luis Borges's Chinese encyclopedia is, of course, amusing because of the contrast between it and the list of Linnaean classification.[24] While the latter seems rational to us, the list Borges presents seems absurd. As Foucault notes, this is not because of the contents of the list per se. Each item on the list names a recognizable group of animals, even if some are fantastic. "The stark impossibility of thinking *that*" arises from the way the list in constructed. Having *et cetera* in the middle, for instance. Juxtapositions deepen the impossibility, with stray dogs following fabulous animals. "What transgresses the boundaries of all possible thought," Foucault writes, "is simply that alphabetical series which links each of those categories to all the others."

> It is not simply the oddity of unusual juxtapositions that we are faced with here. We are all familiar with the disconcerting effect of the proximity of extremes, or, quite simply, with the sudden vicinity of things that have no relation to each other; the mere act of enumeration that heaps them all together has a power of enchantment all its own: "I am no longer hungry," Eusthenes said. "until the morrow, safe from my saliva all the following shall be: Aspics, Acalephs, Acanthocephalates, Amoebocytes, Ammonites, Axolotls, Amblystomas, Aphilisions, Anacondas, Ascarids, Amphisbaenas, Angleworms, Amphipods, Anaerobes, Annelids, Anthozoans . . ." But all these worms and snakes, all these creatures redolent of decay and slime are slithering, like the syllables which designate them, in Eusthenes' saliva: that is where they all have their common locus, like the umbrella and the sewing machine on the operating table; startling though their propinquity may be, it is nevertheless warranted by that and by that in, by that on whose solidity provides proof of the possibility of juxtaposition. It was certainly improbable that arachnids, ammonites, and annelids should one day mingle on Eusthenes' tongue, but, after all, that welcoming and voracious mouth certainly provided them with a feasible lodging, a roof under which to coexist.
>
> **MICHEL FOUCAULT**[25]

A final level of absurdity in Borges's list, and perhaps the most important one for us here, is the observation that there is no *site*, no *place* in which the contents of the list can actually appear as juxtaposed. The list's spatial ordering is purely an ordering of language. Foucault refers to this *site*, following André Breton's use of a phrase from the poet Comte de Lautréamont, as an "operating table," evoking both a material table upon which things might be juxtaposed and the kind of table that might spatially represent a list (such as the periodic table). These animals are never going to be experienced in the same place, nor are they going to be spatially represented in any other form than that of a list of incongruous things.

The lists produced by writers confronting Maxwell Street most certainly do have a *site*. Indeed, it is the site of these lists—the place that was Maxwell Street—that journalists, novelists, and others were trying to capture in their list-making. What they share with Borges's list is the excitement caused by unlikely juxtaposition.

The Senses

List-making provides evidence of both the cluttered thing-rich atmosphere of Maxwell Street and writers' inability to embrace it all. The lists are gestures of futility. They do point, however, to the excessive and dizzying experience the market provided. It was a space that overflowed with sensory experience. Place, especially when seen through the lens of landscape, is often represented as a visual object. Indeed, those who have sought to create places often do so through the production of the visual.

> Places can be made visible by any number of means: rivalry or conflict with other places, visual prominence, and the evocative power of art, architecture, ceremonials and rites. Human places become vividly real through dramatization. Identity of place is achieved by dramatizing the aspirations, needs and functional rhythms of personal and group life.
> **YI-FU TUAN**[26]

Places are experienced, and experience involves all of the senses. We hear, smell, taste, feel, and see them.[27]

Sound

While the eye was dazzled by the array of people and things at the market, it was not only visually that the materiality of the street impressed observers. Taste, sound, and particularly smell were at least as prominent in accounts of the *experience* of Maxwell Street. Stall-holders' cries, music, and other sounds mixed with the smells of the various foods associated with the immigrants who ran the market. In de Beauvoir's account, sound plays a particularly prominent role: "drugstore products are sold amid the clamour of four or five radios," the amplified pitch of a "charlatan" touting a supposed cure-all, "the chatter of thousands of brightly feathered birds." Such attention to the sounds of Maxwell Street recurs throughout a hundred years of journalists' writing about the site.[28]

> The organization of human space in uniquely dependent on sight. Other senses expand and enrich visual space. Thus, sound enlarges one's spatial awareness to include areas behind the head that cannot be seen. More important, sound dramatizes spatial experience. Soundless space feels calm and lifeless despite the visible flow of activity in it, as in watching events through

> binoculars or on the television screen with the sound turned off, or being in a city muffled in a fresh blanket of snow.
>
> **YI-FU TUAN**[29]

The repeated descriptions of the cacophony of Maxwell Street affirm the relationship between sound and the singularity of this particular place. Sound was one of the ways in which Maxwell Street overflowed. It was part of the excess.

The anthropologist Steven Feld has spent his life experimenting with the recording of sounds in a variety of settings. He calls this practice "acoustemology"—acoustic epistemology—and uses it to focus on the role of sound in the process of knowing the world. Place is central to this practice. "The experience of place," he argues, "can always be grounded in an acoustic dimension."[30]

During my visits to Maxwell and Halsted there was little in the soundscape to remark on. Portfolios of sound-in-place change over time, as what were signifiers of urban vitality become noises, sounds deemed out-of-place. The kinds of sounds that made Maxwell Street what is was in, say, 1947, when de Beauvoir visited, are no longer there or, if they do appear, are categorized as nuisance noises—disruptions to be disciplined through the imposition of acceptable sound—a kind or urban muzak.

> If today, economic growth and increasing mobility, and hence more noise, are seen as belonging to one and the same process, a century ago, the Austrian ethnologist Michael Haberlandt consciously differentiated between the sounds of work and of mobility. He could tolerate "the resonance of work," such as the "song of the hammer, the shriek of the saw, the beat and clatter of the workshop, the stamping of machines," remarking that "we live on that money, don't we?" Yet he lamented the "deafening, enraging noise of the alley," which he described as "a mix of the rattle of carriages, of ringing and whistling, of the barking of dogs and the ding dong of bells with a hundred indescribable overtones that submerge in the uproar."
>
> **KARIN BIJSTERVALD**[31]

Frequently, references to sound appear alongside references to smell—the other prominent sense in the market. And as with visible "things," smells, sounds, and activities frequently appear in lists.

> Barkers, spielers, pitchmen, and hucksters shout their wares while radios boom and customers haggle in a dozen languages. Merchandise drapes from awnings, spills over sidewalk stands and creaking pushcarts, litters the pavement and walks wherever the hawkers elect to take their stand. There is the sharp odor of garlic, sizzling redhots, spoiling fruit, aging cheese, and strong suspect smell of pickled fish. Everything blends like the dazzling excitement of a merry-go-round.
>
> ***CHICAGO SUNDAY TRIBUNE*, 1947**[32]

Some lists appear to take on lives of their own. For example, sounds and smells from the preceding example are repeated, with only a slight change in wording, four years later. It appears that the latter journalist visited the archives as well as the market itself. Writing repeats itself.

> It's a throwback to the Tower of Babel with the haggling over process being carried on in a dozen tongues. Barkers, pitchmen and spielers shout their wares over the blare of radios perched on windowsills of dirty, tired-looking buildings elbowing each other like the stands on the street.
>
> The screech of live chickens adds to the din of organized confusion....
>
> From a side street comes the high-pitched voice of a revivalist, wailing a hymn into a microphone. A man sits cross-legged on the sidewalk—adjusting the dials for the loud-speaker while swaying in rhythm to the song.
>
> Maxwell is a street of a thousand smells. The pungent odor of garlic, sizzling hot-dogs, spoiling fruit, aging cheese and pickled fish blend in a unique aroma.
>
> ***CHICAGO SUN-TIMES*, 1951**[33]

The market was frequently referred to as a "bazaar"—a reference that displaced Maxwell Street from the American Midwest to Baghdad or Marrakech. The surfeit of noise and smell, in particular, suggested to observing journalists a distinctly foreign destination. The descriptions of excess, along with the references to places far away, are similar to those in travel narratives for and by tourists.

Waste/Smell

Tourists, of course, and particularly the kind of tourist referred to as "traveler," mixed their desire with disgust.

> Waste is simultaneously divine and satanic. It is the midwife of all creation—and its most formidable obstacle. Waste is sublime: a unique blend of attraction and repulsion arousing an equally unique mixture of awe and fear.
>
> **ZYGMUNT BAUMAN**[34]

In accounts of Maxwell Street, vision and smell combine in a particularly telling way in repeated references to garbage and waste. Indeed, the history of representations of the market swirls with references to garbage, to waste, to noxious smells and cacophonous noise.

> Maxwell Street is ankle deep in dirt and refuse, the buildings that line it are dilapidated and tottering. Each passing breeze wafts a rich emulsion of the odors of poultry, crates, day-before-yesterday's spinach, fresh fish, garlic, cheese and sausage; and along with this emulsion it wafts particles of solid matter and deposits them impartially over the counters piled high with prunes, gum-drops, rye bread, dill pickles and other kosher delicacies.
>
> ***CHICAGO DAILY NEWS*, 1928**[35]

> Behind a delicatessen, a sign in the alley cautioned against dumping. Tea barrels overflowed with garbage, and the way was littered with cartons, paper,

FIGURE 5.
Trash. *Photo by Nathan Lerner. ICHi 2000 194.13 f.3. Courtesy Chicago History Museum.*

broken bottles and refuse, a condition common to Maxwell St. and its environs, where ordinances relating to sanitary and fire hazards may as well be written in Arabic.
***CHICAGO TRIBUNE*, 1957**[36]

As time passed and the market became smaller and its existence more tenuous, references to waste and dilapidation became even more prominent.

The market area is smaller, about three blocks long. The crowds are smaller, the area shows signs of deterioration and crime rate has risen. Vacant lots along Maxwell Street are littered with debris, bottles and boxes.
***CHICAGO TRIBUNE*, 1970**[37]

On Sunday, Maxwell Street will be filled with the smells of cooking food, the steamy smells of hot dogs and the sharp tang of frying onions. But on this day it just smelled of the sour garbage that lay uncollected in the gutter.
***CHICAGO SUN-TIMES*, 1977**[38]

During the week, the dusty vacant lots are more desolate than ever. Shabbily dressed old men sit silently on crumbling stoops and drink wine in garbage-

> strewn alleys; the few remaining buildings sag wearily, burned out stairwells and boarded-up windows telling the perennial urban story of neglect and decay.
>
> ***CHICAGO READER*, 1988**[39]

> Maxwell Street is Chicago's Cannery Row, a Casbah with a frayed fez. Haunt of the homeless and of history, it sits amid garbage heaps in the shadow of Chicago's bursting Super Loop.
>
> ***CHICAGO SUN-TIMES*, 1989**[40]

Maxwell Street was a site where the value of objects was negotiated. Some objects were granted an extension on their life as objects. Others moved gradually back from objects to mere things. Writers who focused on waste were making a judgment about the relative lack of value of many of the things of Maxwell Street. The place became a place for waste, where marginal things became linked to marginal people—the immigrant populations who inhabited the space.

> As long as it is considered as such, waste is cast out of any order and system of values. There is no relation of any kind to history, culture or memory. Waste is worthless, beyond, aside, and even against culture. Waste tends to be chaotic, unstructured, repellent, or even toxic. The only attention that waste, garbage, or rubbish garner is during their destruction or hygienically safe disposal.
>
> **SUSANNE HAUSER**[41]

Throughout this history of accounts of Maxwell Street, we are presented with visions of excess. The people are excessive, the things are excessive, the sounds and smells mingle promiscuously. Waste adds to this sense of excess. Garbage, pollution, and filth underline the status of the area as an immigrant space.

Dirt, as Mary Douglas has famously suggested, is "matter out of place."[42] Smell, too, is often unwelcome in modern spaces except as carefully controlled perfume. The sense of smell has been relegated to the bottom of a hierarchy of senses in modern life.[43]

> That there are bad sounds need not diminish the glory of hearing. That there are delightful fragrances has done little to elevate smell: traditionally, the best odor is not a good odor, but no odor at all.
>
> **DAVID TROTTER**[44]

Descriptions of bad odors are particularly prevalent in the extensive nineteenth-century literature on urban reform, where smell was used to underline the otherness of the urban poor of London or Paris. Similarly, in accounts of Maxwell Street garbage, smell, and sound mingle in uncontrolled ways. These sensations are linked to groups of people such as the homeless and, again, to the bazaar—to sites like Fez or Baghdad famously associated with such a barrage of sensation.

A bad smell can create anxiety because "it comes back from outside, from elsewhere."[45]

> Aromas seem to escape our cognitive consciousness. They belong to a realm of "peripheral" psychomotorial actions, an insistent substrate of incessant movement that makes up so much of what we are, but which we so often choose not to register as thought, even though the stamp of the impressions of this movement constantly influences us. They are part of the landscape of the body which we have so often tried to suppress.
> **NIGEL THRIFT**[46]

Smell, odor, aroma—these are elements of a world that lies beyond what was being constructed as "normal" in twentieth-century America. Writers' references to smell mark Maxwell Street as unregulated and chaotic—as "another space," irreducibly different, where things that are normally suppressed become ubiquitous. Smell points to the "excess" that marked the market.

The excessive is that which needs to be rejected, regulated, excluded, removed. At the same time, it is fascinating, desirable, and exotic. Excess is not possible without order. As Zygmunt Bauman has argued, waste is a by-product of order-making.[47] The more one is concerned with order, the more noticeable is that which exceeds it. In Bauman's formulation, the order-making processes of modernity produce people-as-waste—the homeless, refugees, migrants. While Maxwell Street was inhabited by migrants who had been the "excess" of other places (Eastern Europe, the American South, Mexico, and so on), it was also a site of excessive things—of literal waste. Things become waste as they reach particular moments in their biographies when they cease to be commodities.[48]

Willard Motley/ Ethnography/ the Chicago School of Sociology

Georges Perec's advice on listing the contents of a place had been anticipated on Maxwell Street by the black novelist Willard Motley, a popular writer in the naturalist tradition. Motley made his way into Chicago's literary circles with the assistance of Maxwell Street denizen Nelson Algren, who was in 1940 working for the left-wing literary magazine the *Anvil*.[49] Motley's first novel, *Knock on Any Door*, was published in 1947 and was an immediate success, being made into a Hollywood film starring Humphrey Bogart in 1949.[50] This novel followed the life of Italian American Nick Romano as he falls from being an altar boy, through reform school, to a life of crime as a "jack-roller,"[51] to his eventual execution. This was followed by *Let No Man Write My Epitaph* (1958), also made into a movie, which followed the similarly doomed life of Romano's son.[52]

Motley rarely included black people as significant characters in his novels and is often referred to as "assimilationist." His work is no longer well known, and he tends to be compared unfavorably to other African American writers such as Richard Wright and James Baldwin. At the time, however, Motley's social circles placed him squarely at the heart of an African American writing community that constituted a "Chicago Renaissance" to rival the widely known Harlem Renaissance.[53]

During the 1930s Motley spent many days in the Maxwell Street area gathering material for his novels. Motley's observations can be found in pocket-size spiral-bound notebooks filled with notes written in pencil while sitting and watching the neighborhood. They are archived at Northern Illinois University in DeKalb.

> Warning!—DANGEROUS MAN AT WORK!
>
> (To be worked over many times, polished, made perfect)
>
> This is an honest book, brutal in its frankness, lewd almost in its delineation. It is not meant for people who hold their hands up to their eyes. Those had better not read on. The seamy side of any city is most intriguing. It may be brutal. It may be lewd (Newspapers full of scandal but these sections [are?] the scandals themselves.) This book is full of hard words, stony statements. What people say will be recorded here. What they do will be stated. Those who don't like oaths had better close this book now. Its first pages are full of them. Those who don't like to know what some people in a big city are like had better find other reading. Chicago is no Pollyanna and this is not a Pollyanna story. I have not censored a single prostitute. I have not [marshaled?] a phrase, cleaned a curse word. You will find the dirt and filth of the big city scattered over these pages. All its hardness and harshness, glitter and tinsel. But also, its beauty, intellect, culture. However again, Warning!—Dangerous man at work! I am going to pull no punches. I have warned you. Proceed at your own risk!
>
> **WILLARD MOTLEY**[54]

As with *Chicago: Confidential!*, marginal areas of the city (Maxwell Street and Skid Row in particular) are encapsulated by their status as "real"—meaning gritty and shocking. Journalists' interest in revealing hidden spaces to a voyeuristic readership crosses over into Motley's practice as a novelist. But Motley's use of the Maxwell Street area as a setting for his novels also speaks to other kinds of writing that had emerged as urban ethnography several decades earlier with the Chicago school of sociology.[55]

Under the leadership of Robert Park, the Chicago school developed a view of the city that centered on the idea of the city as an "ecology" with a set of distinct habitats associated with particular groups of people. Central to these urban ecological models were notions of order and disorder.[56] While Park, with Ernest Burgess, developed general models of ordered urban form, his students were sent out to conduct detailed qualitative accounts of city life where it was perceived to be "disordered." These monographs involved long-term fieldwork observing and participating in different subcultures (gangs, thieves, ghetto dwellers, hobos, etc.) that made up city life. This was a key moment in the invention of modern sociology as a discipline and of the research method that became known as "ethnography" and, more precisely, "participant observation."

Motley's archives include an extract from Harvey Zorbaugh's classic Chicago school monograph *The Gold Coast and the Slum*.[57] There was a strong relationship between black realist literature in Chicago and the sociologists at the University of Chicago, who were well known to each other.[58] Their writing strategies also overlapped, with the ethnographic monographs of the sociologists exhibiting a literary quality in their exposure of the lowly undersides of the city and the novels of Motley and others evincing a substantial amount of research and observation. It is illuminating to read Motley's novels alongside the classic sociological monograph, Clifford Shaw's *The Jack-Roller*, which covers similar territory and reaches, albeit in a different style, essentially the same conclusions.[59]

Motley's diaries reveal his belief that Maxwell Street provided particularly rich material for a novelist. He recorded his visits to the area. On New Year's Day in 1941 his diary notes:

> Tramps down West Madison, Halsted Street, Peoria, Newberry, Maxwell were wonderful. Just walking and looking. Some of these walks were with Sandy. Impressions came in from everywhere. The neighborhood is a storehouse for a writer.
>
> This one thing I learned last year, probably the most important thing that any writer who wants to get to the top must discover—
>
> Some authors write at a great distance from their subjects, some very close to their subject. I want to write as a part of my subject.
>
> **WILLARD MOTLEY**[60]

Motley's desire to work ethnographically is clear. He does not want to simply observe—he wants to participate in a place and its life. The practice he advocates would be called, in a different time and place, participant observation. By the end of the year Motley had an apartment close to the corner of Halsted and Maxwell. At 41°51'53" N, 87°38'49" W.

On December 12 he wrote:

> I am sitting at my desk in the Loft writing this—the desk faces the window and out the window the view is very beautiful. Snow is falling. A mere flaking. I can see Maxwell Street, the end of a hot dog stand lighted up and Halsted Street going past with now and then an automobile enlivening it. And a neon sign says BUDWEISER. There is a knock-kneed pushcart with a slant-wise canvas top near a curbstone. It is black against the snow.
>
> **WILLARD MOTLEY**[61]

Motley's account of his methods for researching his novels sounds like a manifesto for ethnography: "I didn't get my information out of books or by casual observation. I didn't go down to Maxwell with my fingers to my nose and a wry disdainful look in my eyes. I went down and lived on Maxwell Street, hobnobbed with the merchants, roomed with the people, drank with them, caroused with them, wallowed in the filth and refuse of the street with them."[62] His notebooks of 1939 provide accounts of drinking sessions in local taverns with Maxwell Street characters whom he interviewed. On June 24, for instance, he asked prostitutes if the cops were tough. Later in the day he "met Matt on location. Sat on a stand and drank ½ gallon of beer a piece. Talked with him about Maxwell Street and learned a lot."[63]

Reading these notebooks, with the fading pencil marks and thumbed edges, gives the impression that Motley was following the instructions for writing place developed by Perec decades later. As with the writing of journalists, Motley makes copious lists of things on the street and things that passed through. "On the stands and sometimes right out on the pavement are dumped all sorts of wares—hardware, toys, trinkets, shoes, hats, clothing." As had numerous journalists, Motley listed as well the noises of the market and the smells of garlic and fish that overwhelmed the senses. "The largest impression is one of a junked-up neighborhood—the stands—basements and lofts filled with nondescript things," he wrote, on another day. "The architecture and structure of the brick building show that this was once a first-class neighborhood," he noted, but his overall impression was that "the buildings kneel to the street." The street was where the action was. What Motley saw, however dizzying, was often surpassed by engagement with his other senses: "in looking for color I found a hell of a lot of odor."

> We become attached to the odorous, flavoursome, colourful place, we establish our dwelling there.
>
> **MICHEL SERRES**[64]

Both *Knock on Any Door* and *Let No Man Write My Epitaph* use long passages describing Maxwell Street to set the scene for a narrative of human failure and decline. In *Knock on Any Door* the street is introduced in terms very like those used by Simone de Beauvoir.

> Nick turned into Maxwell Street. Before him stretched the Maxwell Street Market extending between low, weather-grimed buildings that knelt to the sidewalk on their sagging foundations. On the sidewalk were long rows of stands set one next to the other as far as he could see. On the stands were dumped anything you wanted to buy: overalls, dresses, trinkets, old clocks, ties, gloves—anything. On what space was left near the curb were pushcarts that could be wheeled away at night. There were still other rough stands—just planks set up across loose-jointed wooden horses: hats for a quarter apiece, vegetables, curtains, pyramid-piled stacks of shoes tied together by their laces—everything. From wooden beams over store fronts, over the ragged awnings, hung overcoats, dresses, suits and aprons waving in the air like pennants. The noises were radios tuned as high as they could go, record-shop Victrolas playing a few circles of a song before being switched to another, men and women shouting their wares in hoarse, rasping voices, Jewish words, Italian words, Polish and Russian words, Spanish, mixed-up English. And once

in a while you heard a chicken cackling or a baby crying. The smells were hot dog, garlic, fish, steam table, cheese, pickle, garbage can, mold and urine smells.

WILLARD MOTLEY[65]

As with the journalists, the writer is confronted by an abundance of things, which instigates the writing of lists. These lists, and an attention to the different ways in which each of the senses is engaged at Maxwell Street, are evident in Motley's notebooks.

On the stands and sometimes right out on the pavement are dumped all sorts of wares—hardware, toys, trinkets, shoes, hats, clothing.

Noises—

Noise of victrolas. Men crying their wares. Women calling to each other. Chickens cackling. Roosters crowing. Children crying.

Smells—

Garlic

Fish

WILLARD MOTLEY[66]

In *Knock on Any Door*, the mixture of things and the assault on the senses—blaring radios and raised voices, the centrality of smell—is woven into an account that reflects on the ethnic diversity of the people who populate this place: Jewish, Italian, Polish, Russian, Spanish. In this way, the things, the senses, and the people become scrambled—pointing to how Maxwell Street exhibits its lowly cosmopolitanism and its multitudinous connections to the wider world. This is a place made through mobilities.

In Motley's work, as in many journalists' accounts, Maxwell Street plays its role as a place marked by its ethnic diversity and difference. Motley does not just list the groups of people who inhabit the area but uses things he has observed to highlight their variety. We can follow these observations from his notebooks into his novels. At one point in the notebooks, for instance, he remarks on the plant life of the neighborhood.

A string of red peppers drying from a second story back porch as if proclaiming—this is Mexico.

Surprising growths of green branching from bottles, old cans, make-shift flower boxes in front room windows and on balconies. Sweet potato plants, Wandering Jews, Ferns, Peppermint.

WILLARD MOTLEY[67]

The connection between small plants and the people who plant them appears clearer in *Let No Man Write My Epitaph*.

> No grass grows, in smooth green aprons, in this neighborhood. But from every window hang pin tails, crocks and mason jars of green growing things. Surprising growths from milk bottles and tomato cans in front-room windows and on fire escapes. Sweet potato plants. Peppermint. Ferns. Wandering Jew. Memories of the grape lands of Italy, the wide valleys of Mexico, the long fields of the South, the olive-twisted shores of Greece, Gipsy and Jew recollections of the sun-warmed, rain-colored growths of many lands. Evidence of a love of the soil and what it bears.
> **WILLARD MOTLEY**[68]

Motley uses things, in this case plants, to produce a richer sense of the heterogeneous nature of city life in an urban "ghetto."

Performance

Place in general is performed.[69] The interplay between the choreographies of people going about their lives and the things (including the relatively fixed landscape) that surround them produces rich and often very particular senses of place. This was certainly the case in Maxwell Street—a place where the profusion of things and the throng of bodies interacted in constantly fascinating ways. Motley's notes on Maxwell Street portray a lively mix of people involved in almost nonstop performance among a jumble of tumble-down structures and juxtaposed things. Maxwell Street provided a weekly display of the carnivalesque.

> Went down to Maxwell to "Festival of Lights"—celebrating new lights on Halsted St. Mayor Kelly made speech and turned on the lights. He said he had lived nearby—he came from the bottom and was still with the bottom. Trained bear. Festival queen in his garish, spangle-sprinkled robe. Great crowd. Roof tops across street lined with boys, mostly Mexicans—Went to roof. Tore up newspapers and tossed it on crowd below. To Maxwell Street where a colored band on the back of a truck played—loud speakers being attached. Crowd ringed a drunken Irishman who was dancing in the street. Then a colored couple did a jitterbug dance as the crowd kept time with their feet and hands. Then a Jewish girl and her partner took over. In the roof of the stands stood the youths of the neighborhood shouting and applauding.
> **WILLARD MOTLEY**[70]

In Motley's accounts, as in a wide variety of other writings, it is the performance of Maxwell Street that stands out as central to its character. This performance certainly involves things and objects as well as human bodies, but the material landscape plays second fiddle. Maxwell Street was performed both in the sense of everyday performance of place (shopping, selling, hanging out) and in the more formal sense of performance as deliberately staged event. It was a setting for musicians, preachers, snake-oil salesmen, magicians, and a host of other semiprofessional performers.

> A man stuffed a rag of a girl, perhaps 14 years old, into a box and thrust 20 swords through it. He was turbaned, his head-dress adorned with cheap jew-

elry. Into his face he had rubbed oils that made it tan. 2 cents to see the girl in the box. . . .

A woman about 40 in orange slacks, red bandana handkerchief on her head, a paper shoe box with 2 crabs in it. She sits on an old stand playing with them with her keys. Wears large circular brass earrings. A man comes along. He is pushing a leg-less man in a wheel chair. Stops, lifts crabs—gives instructions on handling them. A crowd gathers. She tells the young man running the hot-dog booth directly ahead—"I got crabs." . . .

Maxwell St. East of Halsted, Sunday. A negro 4-piece orchestra. 3 guitars and a horn. Black men. Sat on pop-case, chair, garbage can at St. curb and played and sang. Money in the bottom of a shoe box. One chorus—"You stole my gal at the dance—and she came home without any pants." Store behind them. Youth in doorway—"Ice cream bars only 3¢. There's plenty of room in the store, come on in everybody." The owner brings out 4 for the orchestra, they eat them. One of them—"they sure are big, almost more than you can eat!"

WILLARD MOTLEY[71]

The Gaze

The early ethnographic tradition of the Chicago school of sociology and the realist style of Motley both represent ways of illuminating parts of the modern city considered marginal or off-limits by those who lived and worked elsewhere. They share with documentary photography an interest in revealing what is hidden to the "respectable" citizen. In this sense, it is possible to talk about an "ethnographic gaze."[72]

The act of observing and describing the "marginal" spaces of the city enacts a process by which the socially marginal become symbolically central.[73] To confirm the normativity of "respectable" society (and, necessarily, respectable space), ethnographers, novelists, journalists, photographers, and others must dedicate significant energy in defining the parts of society (and their spaces) that are not respectable.

> It is a peculiar sensation, this double-consciousness, this sense of always looking at one's self through the eyes of others, of measuring one's soul by the tape of a world that looks on in amused contempt and pity. One ever feels his two-ness,—an American, a Negro; two souls, two thoughts, two unreconciled strivings; two warring ideals in one dark body, whose dogged strength alone keeps it from being torn asunder.
>
> **W. E. B. DU BOIS**[74]

In *Ways of Seeing*, a seminal book on visual culture, John Berger reflects on the way that images of women in Western culture are constructed from a point of view or perspective that is most often "masculine." "Men look at women," he wrote. "Women watch themselves being looked at."[75] The way in which ways of looking have produced effects of masculine power was developed by Laura Mulvey in her work on how women have often been represented in the history of cinema as objects of a "masculine gaze."[76] Since then the gaze has been subjected to a veritable cottage industry of theoretical exposition concerning the power that derives from, and gives authority to, the act of looking.[77]

Michel Foucault famously placed the gaze at the center of his account of disciplinary power. We become disciplined, he argued, through a process of policing the self as if we are being always watched. Here the model of the panopticon, developed by Jeremy Bentham, serves as the central metaphor for the ways in which watching and being watched is activated in a variety of social situations.[78] But there have also been challenges to the construction of the gaze as an overwhelmingly one-way act of power. Notions of a "counter-gaze" and a "feminist gaze" are among these responses, suggesting possibilities for resistance in the act of looking.[79]

The Ethnographic Gaze

The "ethnographic gaze" is a way of looking usually practiced by "us" in observing "them," a way of conceiving of, and then representing, the "other." The history of ethnography (like the histories of photography and other ways of looking) is littered with examples of the relatively powerful looking upon the relatively disempowered—an act of looking that is itself part of that play of power. Thus, for the most part, Chicago ethnographies from around the time Motley was exploring Maxwell Street were undertaken by white, male, middle-class members of Robert Park's sociology program and focused on populations and spaces deemed to be deviant or marginal—gangs, hobos, the slum, jack-rollers, prostitutes.[80] Only rarely (and this is still the case) did ethnography involve the relatively disadvantaged entering the worlds of the elite. Indeed, the Chicago school of sociology was implicated in excluding a number of forms of sociological investigation that existed on the margins of what it counted as "proper" research, including work by anarchist activists, women, and black organic intellectuals.[81] Willard Motley, as a black novelist associated with radical magazines such as the *Anvil*, was very much on the margins of what might count as respectable sociological knowledge, despite the fact that his methods were distinctly ethnographic.

While Motley was not officially sanctioned as an ethnographer, the way he sat and watched Maxwell Street certainly constituted a gaze of sorts. How to assess this gaze is complicated by his position as a black outsider in relation to the scientific and literary world but an insider relative to the area—a "native," and as much a participant as an observer. His novelistic gaze was therefore distinct from that of the educated white outsiders exploring what they considered to be the dark underbelly of the city. Maxwell Street was, at least in part, Motley's world. The entangled and ambiguous politics of different forms of gaze are very much a subject of both his notebooks and his novels.

Rubber-necking

This is particularly the case in relation to the tourists who would pass through Maxwell Street in buses. Motley describes such tourists as "rubberneckers," a word that arose in the 1880s to describe voyeuristic gawkers who stretched and bent their necks to gaze at the world outside.

The tourist gaze is directed to features of landscape and townscape which separate them off from everyday experience. Such aspects are viewed because they are taken to be in some sense out of the ordinary. The viewing of such tourist sights often involves different forms of social patterning, with a much greater sensitivity to visual elements of landscape or townscape than normally found in everyday life.
JOHN URRY[82]

Rubbernecking was, in the early twentieth century, a new automobility-generated version of this tourist gaze.

The rubberneckers visiting Maxwell Street were evidently searching for a very different kind of scene than that provided by well-known sites such as Navy Pier or the Lincoln Park Zoo. They were looking for sites marked by gritty urban "reality," where life was led differently. In this sense, they shared much with journalists, ethnographers, and Motley. The scene included the "vibrant color" of the market but also the conditions and practices of the people who lived and worked there.

By 1909, so-called rubberneck automobiles, accompanied by a "megaphone man," who provided a commentary on the urban landscape, would take the curious spectator on a tour through Chinatown, which included visits to a joss house, a theater, and a restaurant.
PETER X. FENG[83]

Sight-seeing buses pass slowly down Maxwell at night. The tourists gawk. There is always a shouted chorus of "Rubberneck" to greet them. Sometimes the youths of the neighborhood collect rotten tomatoes and other vegetables and fruit and redden the windows of the buses. . . .

My contention is that the sight-seeing tourists who drove down the street in sight-seeing buses are contemptible busy-bodies. They crank their necks and stare. These people don't live down here because they want to but because they have to. Can't do any better and they don't go into the sight-seers' neighborhood and rubber-neck.
WILLARD MOTLEY[84]

There are rubbernecking scenes in both of Motley's Maxwell Street novels. He uses such incidents to reflect in more depth on the politics of looking.

As they walked away the sightseeing bus turned off Halsted on to Maxwell. Slowly the rubberneck bus groaned along Maxwell Street, lights out. The windows were open a little at the bottom. They were white with the powdered faces of women.

The rubberneck bus rumbled slowly, its brakes hissed softly, its balloon tires eased the sightseers over the broken Maxwell Street pavement. The spieler has drama in his gestures, in his voice. Maxwell Street!—The Ghetto!—Chicago's gaudy, picturesque outdoor market. Over here you get odor along

> with your color. . . . The circus barker's voice chuckled. The women stared. The men looked out beyond their pressed and expensive suits.
>
> Max said, "You better give me another slug out of that bottle."
>
> The voices at the windows say, talking to themselves: How interesting! . . . What an ugly, dirty street! . . . How interesting! . . . There's no excuse for being that poor and that dirty. They make my skin creep!
>
> The eyes of slum people come up to the darkened windows of the rubberneck bus, narrow, harden.
>
> A man yells, loudly, "Rubbernecks!" Angrily. Contemptuously.
>
> Charlie, selling hamburgers, says, "Goddam them! They stare at you like they're looking at pigs. I wonder how they'd like it if we went over where they live and stared at them?"
>
> **WILLARD MOTLEY**[85]

These episodes in Motley's notes and novels reveal a complicated interweaving of visions. He is looking at the market and taking notes in his notebooks. The rubberneckers are coming to vicariously thrill at the dark side of the city. Motley observes the rubberneckers. Readers of Motley's books "observe" Maxwell Street in a more removed way that again partakes of a vicarious pleasure in encountering the socially marginal. It is difficult, in this complicated web of visions, to straightforwardly assign power or lack of power, blame or praise. What it does show is that Maxwell Street was a site in which looking (alongside smelling and listening) was just one of many practices that made the street a space of lively encounter.

Vivian Maier

One Sunday in 1967 Vivian Maier, a nanny working in the Chicago suburb of Highland Park, left the basement bedroom that had been set up for her by her employers and headed south on the twenty-five-mile journey into the city. This was not unusual. On Thursdays and Sundays she had the day off and would go into the city to take photographs. On other days she would take one of the children she looked after with her. Maier had converted part of the bedroom she lived in into a darkroom. She took thousands of photographs, most of which never got further than the negative stage. Many failed to get even that far. These negatives, along with canisters of undeveloped film, were discovered is 2007, and her images began to appear on the internet in 2008. Among the negatives were the record of the day in 1967 when she left the comfortable suburb and made her way downtown to, among other places, Chicago's Skid Row, on Madison Street, and then Maxwell Street.

Maier must have been a strange sight with her short masculine haircut and old-fashioned plain clothes. She would walk for miles, occasionally switching accents and making up names. By the end of her life in 2009 she had accumulated some two hundred boxes of photographic material, as well as newspapers and other "stuff"—more than 150,000 images—which were piled to shoulder height in the attic room she lived in in her later life. The room was so full that the floor sagged, and a reinforcing I-beam had to be installed. There is a satisfying symmetry between the hoarding instincts of Maier and

FIGURE 6.
The chicken man, 1967, Maxwell Street Market and Skid Row on West Madison Street in Chicago. Photo by Vivian Maier. Image ownership currently subject to legal dispute.

the piles of things to be found at the Maxwell Street Market on Sundays. It does not seem surprising that she liked to visit.[86]

In 2013 I was working in the archives of the Chicago History Museum, looking through its Maxwell Street files. I had scheduled a meeting with the photographer Ron Gordon. I wanted to use a photo he had taken of hubcaps and wheels at Maxwell Street, and I needed to get a good copy from him and pay for permission to use it. I met him outside the museum, where he pulled up in his car. While we conducted our transaction he told me that he had been part of a team printing images from this amazing collection of recently found negatives by a nanny called Vivian Maier, and that she too had visited Maxwell Street. There was, in fact, an exhibition of her images in the lobby of this very museum—an exhibition I had ignored many times on my way up the stairs to the archives room. Our business complete, I returned to the museum and walked around the exhibition. There were enormous blow-ups of some of her photographs as well as some smaller images, arranged in a sequence that followed one of her photographic walks. The sequence ended in

Maxwell Street.[87] Among her Maxwell Street pictures were images of market bric-a-brac, weathered faces, children looking nonchalant, street performers, and a man playing accordion with a chicken on his head.

In Maier's photograph, we see the chicken man in a space created by the bodies of the people at the front of the group of spectators. The photographer and the viewer of the image are positioned as spectators—as passersby whose progress through the market is halted by the magnetic pull of performance. The image is taken from a low vantage point, as if the photographer were one of the children who often accompanied Maier on her urban drifts. There is a sense of shyness, as if Maier were hiding in the crowd. The focus on the chicken man and his performance reflects many images of Maxwell Street through the decades. Most other photographers, however, would have pushed through the spectators to get a clearer shot. This image is uncertain. The contrast in the image makes silhouettes of the spectators who frame the chicken man standing against the wall of Leavitt's Deli on the northwest corner of the intersection of West Maxwell and South Halsted Street.

41°51′53″ N, 87°38′49″ W

Photography

Markets are magnets for photographers. During my frequent visits, over the course of a decade, to the displaced Maxwell Street Market—now on Desplaines Street, across a major highway and several blocks away from what is left of the redeveloped Maxwell Street—I take endless photographs. I see a good number of other people also taking photos, some with their mobile phones, some with the complicated paraphernalia of tripods and multiple lenses. There is something about this market, and markets in general, that makes people want to take pictures. My images, those of a rank amateur, focus on the profusion of things—brightly colored Catholic kitsch, boxes of chilis curled around each other, or the unlikely juxtaposition of a fur, a sombrero, and a *Rocky* album.

Markets have been a constant in the history of documentary and street photography. The French photographer Eugéne Atget, one of the original "street photographers," would wander the streets of Paris documenting the streets and buildings that had not been knocked down to make way for the wide boulevards and grand monuments of Georges-Eugène Haussmann's remodeled city.[88] Between the 1880s and the 1920s he photographed many aspects of the city that remained invisible to the casual observer. Among his favorite subjects were the rickety stalls of vendors in the back alleys of Paris. Most often the photos have no people in them, reflecting Atget's practice of arriving early in the morning with his old-fashioned equipment that could not capture anything that moved too much.

> Not for nothing have Atget's photographs been likened to the scene of a crime. But is not every square inch of our cities the scene of a crime?
> **WALTER BENJAMIN**[89]

FIGURE 7.
Rocky. *Photo by author.*

In the 1930s László Moholy-Nagy, a photographer/artist who formed the New Bauhaus in Chicago, visited London and found himself moving away from his formal images of pattern and forms as he was confronted with the street markets of London.

> The Photographer can scarcely find a more fascinating task than that of providing a pictorial record of modern city life. London's street markets present him with an opportunity of this kind. It is not, however, a task to which the purely aesthetic principle of pictorial composition—which many readers may expect in my work—can be applied, for from its very nature it requires the use of the pictorial sequence and thus of a more effective technique approximating to that of the film. I am convinced that the days of the merely "beauti-

FIGURE 8.
Boutique de fruites et légumes, Rue Mouffetard, *1925. Photo by Eugéne Atget. Gilman Collection, Purchase, Ann Tenenbaum and Thomas H. Lee Gift, 2005, Metropolitan Museum of Art.*

ful" photograph are numbered and that we shall be increasingly interested in providing a truthful record of objectively determined fact.

To many peoples' minds the street market still suggests romantic notions of showmen, unorganised trade, bargains and the sale of stolen goods. The photographic report can either encourage or correct these ideas. I consider the latter to be the more important task, since in my opinion these markets are primarily to be regarded as a social necessity, the shopping-centres in fact, for a large part of the working-class.

LÁSZLÓ MOHOLY-NAGY[90]

Exploring this entanglement of the material, the performative, and the representational involves navigating a complicated labyrinth of ways in which Maxwell Street became a place for the practice of looking.

Photography/Representation/Practice

Representation is often regarded as a moment of deadening. The moment when a text is written or a photograph taken is, in this interpretation, a moment when the world is made to stop. Thereafter the scene is (re)presented as if it were always as it had been at that moment. Forever. Representation, here, is posited as the enemy of the practical and performative—the lively ongoing nature of the world.[91]

At Maxwell Street photography was a form of practice located in place. Representation is "doing" as much as any other form of ongoing engagement with the world. The act of taking a photograph is active. The content of the image does work. The viewer of photographs, too, is far from a passive receptacle of information. There is liveliness everywhere you look in the act of representation.[92]

> Can we not say that there was already in photography, in the classic sense, as much production as recording of images, as much act as gaze, as much performative event as passive archivization?
>
> **JACQUES DERRIDA**[93]

The liveliness of the representational act of photography can be informed by two seemingly contradictory impulses. On the one hand, there often appears to be something internal to the photograph that catches our attention and draws us into lively engagement. In *Camera Lucida* Roland Barthes argues for an indefinable "singularity" that often determines our relationship to an image.[94] The nature of photography, he argues, rests on the fact that it can reproduce moments that are unrepeatable. A photograph, then, is an "event"—a moment that cannot be transcended or made to speak for something else, elsewhere. Barthes does not deny that there are elements of looking at a photograph that refer to a wider set of expectations and lead us to acts of interpretation. This aspect of the image he calls its *stadium.*

More important, however, is Barthes's notion of the *punctum*—a moment of singular drama in an image that connects on an intensely personal level. The *punctum* creates a connection with the viewer, Barthes argues, that is beyond language. It is an element of an image that reaches out and "wounds" the observer (the word *punctum* is derived from the Latin word for trauma). The *punctum* appears to be a nonrepresentational element of this form of representation where no amount of reference to "context" is likely to help in interpretation.

But perhaps even the *punctum* of an image varies with the time and place of its reception.

> The photograph is an "incomplete" utterance, a message that depends on some external matrix of conditions and presuppositions for its readability. That is, the meaning of any photographic message is necessarily context de-

termined. We might formulate this position as follows: a photograph communicates by means of its association with some hidden, or implicit text; it is this text, or system of hidden linguistic proportions, that carries the photograph into the domain of readability.

ALLAN SEKULA[95]

A photograph is not sufficient unto itself. When we see an image of African American children playing on Maxwell Street we bring to that act of viewing a matrix of previous experience, including our familiarity with many other images of children on streets.

Adequately understanding a photograph, whether it is taken by a Corsican peasant, a petit-bourgeois from Bologna or a Parisian professional means not only recovering the meanings that it proclaims, that is, to a certain extent, the explicit intentions of the photographer; it also means deciphering the surplus of meaning which it betrays by being a part of the symbolism of an age, a class or an artistic group.

PIERRE BOURDIEU[96]

This argument could be seen as a case of the "deadening" effects of representation—diminishing the power of the image to surprise and create an "event" out of viewing.[97] On the other hand, it may be that the context within which we approach a photograph makes our viewing of it even more lively. It allows the singular photograph to make connections to multiple (and ultimately undetermined) worlds beyond the frame. A photograph's eventfulness may be memorable and surprising because of (not despite) its contextual character.

Looking at photographs in this way simultaneously denaturalizes the image and increases its layers of meaning and thus its capacity for surprise. It also takes the photograph out of the realm of the taken-for-granted and increases its capacity to connect to the world outside of its immediate presence. We are alerted to the fact that the language of photography is not entirely internal and cannot be satisfactorily interpreted in formalist ways that rely on meaning intrinsic to the image. Photographs are not transparent (re)presentations of some external "reality" but are instead implicated in the production of that reality. They are alive and active. They simultaneously record and make the world.

What is the relationship between photography and place? More precisely, in what ways does photography feed off and into kinds of places that become privileged sites of representational practice? Why do photographers return, again and again, to Maxwell Street?

Photographs are not simply records of place but ways of engaging with them. "Photography," Joan Schwartz and James Ryan write, "remains a powerful tool in our engagement with the world around us. Through photographs, we see, we remember, we imagine: we 'picture place.'" Photographs have been key tools in the creation of imaginative geographies that do active work in the places that are pictured. They shape perceptions and attitudes. And the work of photography does not just happen in the spaces of the place that is photographed and in the photograph itself but in "the spaces of photographic practice, reproduction and circulation."[98]

Photography and Marginal Place

> The rise of the modern city coincided so closely with the development of photography and other mass printing technologies that the "urban" and the "visual" became deeply, inextricably entwined.
> **JOSEPH HEATHCOTT AND ANGELA DIETZ**[99]

Maxwell Street is a marginal place, the place of the other, a place far removed from the lives of "respectable" middle-class viewers of photography. Places such as this become rich sources of imagery for a bourgeois imagination that feeds off the slightly illicit excitement that comes from being exposed to a little bit of the "other." Maxwell Street was such a place—a market and a "slum"—a place that, by its very existence, confirmed the "respectable" nature of the distanced viewer.

The spatiality of vicarious viewing maps onto the necessary relationships between spaces defined as "slums" or "ghettos" and the spaces from which they are viewed. There is more than one way in which value is extracted and exported.

> A process of displacement and what I call "accumulation by dispossession" lie at the core of urbanization under capitalism. It is the mirror-image of capital absorption through urban redevelopment, and is giving rise to numerous conflicts over the capture of valuable land from low-income populations that may have lived there for many years.
> **DAVID HARVEY**[100]

Maxwell Street Market was a place of mobility. The local people were immigrants, and shoppers came in from all over Chicago, along with rubbernecking tourists. The objects traveled too. But this was a place where everything clashed and bumped into each other. People were forced to communicate and haggle. While there were signs along the streets, they did not dominate life. Prices were often not advertised but had to be arrived at. The entangled life of people and things had to be negotiated, often noisily and with gusto. Maxwell Street was a hyperplace of enforced communication.

If, as Ralph Waldo Emerson wrote, "Cities give us collision," then Maxwell Street was the epitome of what it is to be a city.[101] It was aspects of this collision that have inspired photographers from the very first years of the twentieth century to visit and revisit the market.

László Moholy-Nagy found the London marketplace to be a particularly rich source of visual imagery because of both collision (a largely horizontal notion) and the layering of complexity (a largely vertical notion) that such a place necessitates.

> The subject is a vast one, comprising problems of history, sociology, economics and town planning. It is approached in this book by means of literary and impressionistic photo-reportage. This method of studying a fragment of present-day reality from a social and economic point of view has a wide general appeal. The text provides considerable opportunities for this study and it was my aim to underline these opportunities through the pictorial record.
> **LÁSZLÓ MOHOLY-NAGY**[102]

The market, as a particularly intense kind of place, of both collision and layering, gathers not just things and people but layers of social, economic, and historical meaning that the photographer can explore. The market also provides several scenarios played out repeatedly for the photographer to capture. Markets became part of a stock set of places that quickly became clichés.

> Thus, for all its ability to provide us with images of everything, to buck the generical constraints that continued to govern painting, the photography of the city seems simply to have become obsessively preoccupied with a new range of visual topics: people leaping puddles, empty chairs, road sweepers, markets, shop windows, café mirrors.
> **CLIVE SCOTT**[103]

The Maxwell Street area was not just a market, however, it was also a "slum" or "ghetto."[104]

Alongside the history of street photography that focuses on the color and activity of market life there is also a documentary tradition that focuses on life on the margins of "respectable society."[105] Almost since photography's inception, photographers have been concerned with the act of documenting parts of life that were deemed to be removed from the experience of people who were likely to view the images. This continued a broader tradition of writing (and illustration) that sought to cast light on the seemingly shadowy world of those parts of the city associated with the poor and the destitute and, by association, the amoral and immoral.

The Slum

The social geography of the slum was a source of almost constant fascination to the inhabitants of the "respectable" world. Investigations of the dark side of urban life in the newly urbanized nations of Europe and then North America were simultaneously reports on the "conditions" of the poor that sought to provide reform and relief, and voyeuristic, often sensational narratives that could be found in the drawing rooms of the better-off. They exhibited elements of both disgust and desire in much the same way as orientalist writing on the "East" or "darkest Africa."[106] Classic examples of such texts include Henry Mayhew's *London Labour and the London Poor*, which was written in the 1840s and included written portraits of various street "types" such as "mudlarks" and "tanners." Forty years later Charles Booth's *Life and Labour of the People in London* ran to nine, and later seventeen, volumes and took the practice of documenting the urban "other" in a more scholarly direction.[107] This formed the starting point for the discipline of sociology, which had one of its high points in Chicago in the early twentieth century with the formation of the Chicago school of sociology.[108] This tradition of social documentary had crossed the Atlantic and made its way to New York City by the 1890s. Possibly the best known of the early accounts of the slums of New York City was Jacob Riis's *How the Other Half Lives*, published in 1890. Riis was a photographer as well as a writer. His prose focused on the infamous tenements of New York City's Lower East Side and Harlem. These buildings, poorly lit, overcrowded, and managed by negligent landlords, are presented as the site of an overwhelming "flood" of immigrants.

> The sea of a mighty population, held in galling fetters, heaves uneasily in the tenements. Once already our city, to which have come the duties and responsibilities of metropolitan greatness before it was able to fairly measure its task, has felt the swell of its relentless flood. If it rise once more, no human power may avail to check it.
>
> **JACOB RIIS[109]**

Riis's exposé of the New York tenements is more notable for the inclusion of his photographs. Riis was not the first documentarian of the "other half" to make use of cameras. In London, the Scottish photographer John Thomson had returned from photographing China and Cambodia to continue his endeavors close to home. *Street Life in London* was published in 1876 and 1877 and featured images of the kinds of people Mayhew has earlier written about—swagmen, beggars, boot-cleaners, and the like.[110] Thomson's images shocked "respectable society" and philanthropists into efforts to ameliorate the conditions of the poor. The images, combined with the texts of social reformers, served both to provide a disciplinary gaze from a lofty height and to titillate through their exoticization of the low and the marginal.[111] It is impossible to consider the history of photography without seeing it as part of an expanded sense of the gaze that included the nascent social sciences.

> Made practicable at a time when vision and knowledge came to be inextricably linked, the photograph offered a means of observing, describing, studying, ordering, classifying and, thereby, knowing the world. The rhetoric of transparency and truth that came to surround the photograph enabled it to take up a position between observer and material reality. There, photographic facts generated meaning, and gave rise to action. There, "photographic seeing" became a surrogate for first-hand observation. There, the photograph served as a site where broader ideas about landscape and identity were negotiated.
>
> **JOAN SCHWARTZ AND JAMES RYAN[112]**

In New York, Riis certainly played a part in the institutionalization of a disciplinary gaze directed at the poor.[113] He worked as a police reporter and wrote for several New York newspapers, reporting back to the living rooms of the well-to-do the activities of the city's underworld. One method he developed in his reporting was the use of a magnesium flash-gun to illuminate the (literally) dark spaces of the slums. He used flash technology to expand the gaze both temporally (into the night) and spatially (into the interiors of slums, its alleyways and courtyards). Many of his images were recorded in basements or crowded interiors. His arrival was not always welcomed.

> It is not too much to say that our party carried terror wherever it went. The flashlight of those days was contained in cartridges fired from a revolver. The spectacle of half a dozen strange men invading a house in the midnight hour armed with big pistols which they shot off recklessly was hardly reassuring, however sugary our speech, and it was not to be wondered at if the tenants bolted through windows and down fire-escapes wherever we went.
> **JACOB RIIS**[114]

Riis presented his images in outdoor lantern slide shows, projecting them onto a sheet strung between trees. His images of the urban poor and their slum homes were meant as both entertainment and education.

Documentary Photography

Riis's photographs are not those of an artist. They claimed, as all "documentary" photography was later to do, to be an account of actuality. They featured the buildings and neighborhoods of New York's most infamous districts. They tended to reveal dark corners and featured a considerable degree of waste and dirt on order to shock viewers. Often the images focused on children (who were particularly dirty) to suggest innocence and goad the conscience. The insides of homes were ramshackle and crowded in a way that contradicted the model of domestic hygiene that was popular at the time. While the images presented the illusion of "reality," many, such as the iconic *Children Sleeping in Mulberry Street*, were posed.

The spaces in Riis's photographs are remarkably empty of people considering he was working in a square mile with more than three hundred thousand residents. The people in his images had to stay relatively still. With the technology at his disposal it was difficult to capture the hustle of the crowded "slum."

What Riis's images do reveal is the tight connection between social research, discipline, and moral discourse about the poor and their city. The camera was a key tool in how the marginal and overlooked parts of the city became symbolically central to the bourgeois imagination. This emerging tradition of urban documentary photography was implicitly linked to a wider emerging documentary tradition of shining a light (metaphorically and literally) on spaces conceived of as dark in the bourgeois imagination. Darkest London or New York were, in this way connected to "darkest" Africa or Asia—spaces in which

FIGURE 9.
Children Sleeping in Mulberry Street, New York City, *1890. Photo by Jacob Riis. Wikicommons https://commons.wikimedia.org/wiki/File:Riischildren.jpg.*

photography and anthropology combined to make the apparently disordered and illegible, intelligible to the inhabitants of the metropolitan center.

Similarly, the early documentary tradition would focus on all manner of marginal people (in marginal spaces), most famously, Native Americans and "Eskimos."[115] In addition to this tradition of documenting the other, photography was quickly enlisted into more deliberate and direct disciplining of bodies deemed worrying to respectable society. Men bearing cameras began to appear in all the most likely sites of discipline, in which the viewer exerts the most extreme form of disciplinary gaze on subjects who are further marginalized and stigmatized in the process.

> Photography was used in the many state institutions that proliferated from the 1870s onwards: hospitals, asylums, orphanages, schools, workhouses, barracks, reformatories, prisons, police stations. In those spaces the criminal, the insane, the hysterical, the nomadic, the orphaned, the immoral, the non-civilised, were photographed as evidence of their deviance.
> **GILLIAN ROSE**[116]

> Archives of photographs with distinct visual genealogies function as sites through which narratives of national belonging and exclusion are produced. . . . In the nineteenth century, state-sponsored institutional archives such as the Rogues' Gallery of criminal offenders and scientific archives of racial others created a normative space that delimited the bounds of "true" national belonging. Such archives marked the limits of white middle-class

> American identity and encouraged constant surveillance of the social body for "deviant" outsiders.
>
> **SHAWN MICHELLE SMITH**[117]

By photographing the urban poor in Maxwell Street and elsewhere, the population of the inner city was connected to other populations in other places, ranging from Native Americans to the insane and criminal. Photography was a practice of spatial connection as much as representation.

The unequal relationship between the documentary photographer and those being photographed is clearly a geographical one. Photographers invariably locate people being photographed in sites that are deemed to be "other"—the marginal, the dangerous, and the mysterious. The "slum" is placed in an imaginative geography that includes the high arctic and the African bush. Value is extracted.

This is not to say that documentary photographers had evil intent. Riis, Lewis Hine, and others believed that by shining a light into the recesses of the city they could raise awareness among bourgeois spectators that would eventually lead to amelioration of the condition of the "other half." Martha Rosler has argued that while this may have been true of the early documentary photographers, it became rapidly less true, as the "conditions" of the poor and marginal were hardly mysteries to most likely observers of the images. Most people in New York City, even when Riis was practicing his photographic raids, knew there were poor people on the Bowery. Rosler suggests that the background discourse that made these images intelligible was a mixture of police surveillance and liberal ideology. Documentary photography, she argues, was part of a disciplinary gaze in which the "victims" of capitalist urban conditions became victims of the photographer too. In this way, documentary photographers were like Motley's rubberneckers.

> Documentary testifies, finally, to the bravery, or (dare we name it?) the manipulativeness and savvy of the photographer, who entered a situation of physical danger, social restrictedness, human decay, or combinations of these and saved us the trouble. Or who, like astronauts, entertained us by showing us the places we never hope to go.
>
> **MARTHA ROSLER**[118]

The relationship between photography and the slum has been most often interpreted as an unequal relationship between the viewer and the viewed, between an active subject and passive objects. This account is premised on a pervasive one-way gaze in which the bourgeois looks upon the poor with a mixture of disgust and desire. The "other half" is both appalling and exciting.

Documentary

The history of photographers in Maxwell Street reveals a complicated and contingent relationship between photography and place. A large collection of images spans more than a century and cuts across a number of photographic genres. Some can straightforwardly be labeled "documentary." Others might fall under the headings of "street photography" or photography self-consciously intended as "art." Many images could be filed as all three. While documentary photography, not surprisingly, has most often been thought of as aiming to "document" a part of the world or scenes related to some social issue, "street photography" has been seen as more artistic, a slightly amused attempt to capture rich moments of life on the street. While the documentary photographer resembles official sources of authority—government, planners, academic discourse, the police—the street photographer is more likely to be positioned as a flâneur—someone just passing through, enjoying the scene. Documentary photography expresses "emotional intensity," while the street photographer tends more toward "amusement."[119]

The archives at the Chicago History Museum contains hundreds of photographs of Maxwell Street Market—from the first decade of the twentieth century to its demise in the 1990s. Many of the photographers are unnamed; some are amateurs, others from the press. Among the known photographers are some of considerable repute. The first images in the first folder (1900–1909) are probably from around 1906, six years before the city officially sanctioned the market. There are few images in folders for the early decades or in the final folder (1970–1999). The 1950s has the most images and is divided into six folders. Many of these were taken by members of the Fort Dearborn Camera Club, one of the oldest continuously running camera clubs in the United States. Students, it appears, were sent out to Maxwell Street as an assignment—to create a photo essay or visual record of this foreign place right on their doorstep.

On one level the photographs provide a factual record of the changing nature of the market. Early images are notable for the unpaved streets and the white, largely Orthodox Jewish population that inhabits them. Several images focus on the wooden buildings that surround the market, often with no people in sight. In the 1930–1939 folder, a small number of African American stallholders and shoppers enter the visual record, and cars begin to cruise the streets. By the 1950s, when members of the Fort Dearborn Camera Club begin to visit, Maxwell Street starts to become a space of artistic representation, with more self-consciously "arty" pictures foregrounding the objects of the market. The 1960–1969 file reveals changes in population—now almost entirely African American or Latinx—and in the general landscape, as dereliction becomes more obvious: buildings appear empty, waste more pervasive. The final folder contains only one image.

Over the course of the century, the contents of the market stalls change from predominantly food (mixed with medicines, suits, and other clothing) to assorted leftovers of life—hubcaps and hardware. But beyond this sort of evidence of a changing place, key themes emerge and cross over between decades.

Elevated Views

The corner of South Halsted and West Maxwell Streets (41°51'53" N, 87°38'49" W) appears frequently in the archive, as photographers attempt to capture an elevated sense of place (see fig. 10). This positioning matches the inherited sense of place as *landscape*—a defined section of land seen from a distant, usually elevated point.[120] These images focus on the heart of the market, the intersection of Halsted and Maxwell—the only intersection of Maxwell Street that remains—the site where Maier found her chicken man and Motley looked out on a snowy night. In contrast to Maier's low angle, the elevated shots attempt to get above the hustle and bustle to find a strategic view—a view that turns the activity of the market into a scene. This enacts scopic power by placing the photographer, and then the viewer, in an authoritative position, at a vantage point above the fray that turns everyday liveliness into a "scene" or a "landscape."

> The totalizing eye imagined by the painters of earlier times lives on in our achievements. The same scopic drive haunts users of architectural productions by materializing today the utopia that yesterday was only painted. The 1370 foot high tower that serves as a prow for Manhattan continues to construct the fiction that creates readers, makes the complexity of the city readable, and immobilizes its opaque mobility in a transparent text.
>
> **MICHEL DE CERTEAU**[121]

The views taken of Maxwell Street from the top floors of buildings are not as elevated as de Certeau's view from the Twin Towers, but they enact some

FIGURE 10. *Bird's-eye view looking west over Halsted and Maxwell Streets, with Vienna Red Hots stand at bottom center. Photo by Mildred Mead. ICHi-24634. Courtesy of Chicago History Museum.*

FIGURE 11.
Aerial view of the Maxwell Street Market looking east from the roof of 733 West Maxwell Street. Photo by Americo Grosso. ICHi-34459. Courtesy of Chicago History Museum.

of the same desires. It is easy to imagine a photographer enveloped by the hustle of the market seeking a safe spot at which to set up a tripod and attempt to capture an overall scene. We are still close enough for the density of the crowds to give a sense of the busyness of a space of market practice, but it is a removed sense. This is one way in which documentary photography works—by creating a sense of distance, rather than engagement, between the observer and the observed. People become landscape.

Street Photography

Street photography (as opposed to photographs of streets) depends on being in the crowd, amid the activity catching the fleeting performance of place. And this perspective necessitates a change in technology. The street, and the market in particular, could not easily be captured by images requiring long exposures in studio settings. The goal of being in the crowd and, as much as possible, not being noticed is what led László Moholy-Nagy to embrace more mobile technologies for his images of London markets in the 1930s.

> Thus after several attempts with a large camera I always returned to the Leica, with which one can work rapidly, unobserved and—even in the London atmosphere, or in interiors—with a reliable degree of precision. I hope, therefore, that many a defect incompatible with the standard of photographic quality I have so often demanded in theory will be condoned by the reader, in view of the rapid and unprepared fixation of lively scenes that could never have been posed.
>
> **LÁSZLÓ MOHOLY-NAGY**[122]

Markets are mobile places that cannot be staged. Market photography demands mobile photographers.

Images of Maxwell Street taken at street level—images of market activity—appear throughout the archives. In the first decades of the twentieth century the shutter speeds available on handheld cameras frequently "failed" to stop the motion of the market. In one of the earliest images (from around 1906) a lone woman sits huddled among dirt and snow selling a few meager products.

This image (fig. 12) has a spectral quality to it. The freezing woman with her few things, the dirty old snow, the line of people in the background waiting outside a fishmonger, and, most of all, the hazy figures of a pedestrian making his way down the middle of the street and a horse's head entering the scene from the right. Camera technology then would not have been able to capture a man (or horse) walking with such purpose. The figure gives the sense of the ephemeral nature of the multiple mobilities of market life that are lost to even the most creative archive. I find this image to be especially poignant as it was this inability to "capture" the practice of the market that would lead to its eventual demise.

Sometimes, as in figure 13, the subjects are clearly aware that they are being photographed and play up for the camera. These stall-holders are real people, people who stare confidently, in an amused way, back at the photographer. They are posing. In the background market activity continues, as the blurred, translucent walking figure testifies. At other times the subjects appear less implicated in the process of photography.

Many of the images in the Maxwell Street archives could comfortably be classified as "street photography"—a genre that takes "the street" as its subject, in images that are usually well-peopled and often taken covertly with relatively unobtrusive equipment.[123] Due to its highly mobile nature, street photography could never be as composed as high art photography was from the 1930s onward. The images of Robert Frank in his 1950s book *The Americans*, for instance, were unfavorably compared to the classical, proportioned beauty of the landscape photographs of Ansel Adams.[124] The magazine *Popular Photography* called the images in the collection, eventually recognized as one of the two or three most influential books of street photography, "meaningless blur,

FIGURE 12.
Street vendor selling notions on Maxwell Street, ca. 1906. Photo by Charles C. Clark. ICHi-66024. Courtesy of Chicago History Museum.

FIGURE 13.
Vendors on Maxwell Street. 1922. Photo by Chicago Daily News. *DN-0075248. Courtesy of Chicago History Museum.*

FIGURE 14.
Man wearing a straw hat and playing harmonica, ca. 1930–1950. Photo by Monty La Montaine. ICHi-51940. Courtesy of Chicago History Museum.

grain, muddy exposures, drunken horizons and general sloppiness."[125] Frank was even called unpatriotic, in that his endeavors failed to show America as conforming to established ideas of beauty. Frank, however, was uninterested in the formal compositions of landscape photography. He preferred to take photographs of "things that move." He also moved a great deal himself—traveling around the United States in a used car for two years to collect his images. Appropriately enough, Jack Kerouac wrote the foreword to his book.

There are, in the Chicago History Museum's Maxwell Street photo archive, no images taken from a car window, but there is a clear impression of photographers mingling with the crowds and walking among the stalls. There is a constant sense of "things that move" in these images.

A good deal of the movement on the street was performative. Some was performance in a general sense—visitors performing the role of shoppers, stall-

FIGURE 15.
Jazz with Junk on Maxwell Street, *November 1959. Photo by Clarence W. Hines. ICHi 12834. Courtesy of Chicago History Museum.*

holders the exaggerated roles of sellers—and some in a more specific sense: there were as well musicians, magicians, preachers, snake-oil salesmen, and con artists. Shoppers and strollers gathered around performers, listening and watching. These moments of performance proved to be a magnet for the photographers who regularly populated the street. Blues musicians were a favorite subject.

Figure 15 shows "Daddy Stovepipe" (real name, Johnny Watson) performing for change on Maxwell Street. Born in 1867, Daddy Stovepipe traveled around the southern United States and Mexico starting around 1900, performing in mariachi bands and in minstrel shows. In the 1930s he performed and recorded with his wife, Mississippi Sarah. Following her death, he returned to the life of an itinerant street musician and made his way to Maxwell Street, where he started performing in 1948. This image was used to produce the bronze statue of a blues player that now appears on what remains of Maxwell Street. He was ninety-two years old when it was taken, and died four years later.[126]

Images of street performance reveal the performers' awareness of the need to be noticed in a large crowd. Eccentric clothing and notable hats are *de rigueur*. Crowds gather round musicians and magicians. Among the crowd were the photographers whose role was to notice—while not being noticed.

FIGURE 16.
Man holding a dollar bill in front of a crowd at the Maxwell Street Market, ca. 1930–1950. Photo by Monty La Montaine. ICHi-67113. Courtesy of Chicago History Museum.

The photographer is part of an audience. Often street photographs include the backs of other passersby who have stopped for a moment to catch some of the action. The performers are notably eccentric and have a "spiel"—they dress up and act out to attract the strolling gaze of the pedestrian. One stands on a raised platform and speaks into a microphone while using a dollar bill as prop. Another wears an "Indian" headdress.

One character who peppers the archive is the man who had a chicken on his head. His name was Anderson Punch, but called himself Casey Jones. To most he was just the "chicken man." He was 104 years old when he died in 1974. Many visitors to Maxwell Street, over four or more decades, remembered encountering the chicken man. He appears, for instance, in the documentary film *And This Is Free*. He is remembered fondly on the blog *Chicago Stories*.

FIGURE 17.
Vendor speaking to two men and a boy at a booth at the Maxwell Street Market, ca. 1930–1950. ICHi-67114. Courtesy of Chicago History Museum.

The old black man with a beard and a cumulus of snow white hair walked along with the rooster atop his ancient, ruined fedora. He would draw a crowd by pulling out his old squeeze box from a battered tin case and playing, the chicken riding on his head the whole time. After the onlookers each put down a dime for the show, the old man took the bird off his head and laid it on the pavement. Covering it with a cloth, he told it to "Go to sleep. Go to sleep." The rooster would lie there silently while he played and kept up a steady patter in a high-pitched, toothless voice, telling how he had trained 37 roosters during his years as a show-man. Then he would remove the cloth. The chicken would wake up, scratch-dancing around the sidewalk to the music. The rapt crowd watched as if hypnotized.
CHICAGO STORIES[127]

The chicken man's entertaining performance, not to mention the chicken perched on his head, drew the attention of photographers. Enter "chicken man" and "Maxwell Street" on Google, and you will find a handful of images from different years as well as some stills from *And This is Free*. He presented himself as a spectacle and largely succeeded. In one image, from 1964, he is pictured on the sidewalk complete with chicken, accordion, and a sign around his neck that states his "name" (Casey Jones), his birth year (1870), and his age (94).

Children

Children feature in a substantial number of images in the Chicago History Museum's image file for Maxwell Street, as they do throughout the history of street photography.

Images of children sleeping on the city streets remind us of Jacob Riis and the role of documentary photography in attempts to improve the lot of the poor. The Maxwell Street photos of children are voyeuristic. Those holding the cameras are more than likely to have been well-to-do, tourists in a "foreign" land where children are part of street life.

These children appear to be without parents or other adults. They seem disconnected from both family life and the kinds of private or organized space where children are supposed to be in respectable society. These children do not seem to be "innocent."

There is a long history of street children as figures of both sympathy and moral panic. The latter arises when children transgress the bounds of bourgeois moral geographies. In the modern Western city, children are expected to be at home, at school, or in some other approved, functional space.

> Mr Hawes carefully follows this theme through the cities of 19th century America. The Common Council of New York City were begged by the Rev. John Stanton in 1812 to "make an attempt to rescue from indolence, vice, and danger, the hundreds of vagrant children and youth, who day and night invade our streets," and in 1826 in Boston, the Rev. Joseph Tuckerman complained about the "hordes of young boys who thronged the streets and at times disrupted the operations of the city market" while by 1849, George W. Mansell, chief of police in New York, reported to the Mayor calling attention to "the constantly increasing numbers of vagrant, idle and vicious children" who swarmed in the public places of the city. "Their numbers are almost incredible . . ."
>
> **COLIN WARD**[128]

As children grow up they increasingly enter public space and range further.[129] The children on the street in the Maxwell Street images are significantly younger than the appropriate age for unsupervised street life. As such, they become signifiers, part of the construction of Maxwell Street as a different kind of place, one that is both picaresque and troubling.

> Since the Enlightenment, this has been one of the mustiest speculations of the pedagogues. Their infatuation with psychology keeps them from perceiving that the world is full of the most unrivaled objects for children's attention and use. And the most specific. For children are particularly fond of haunting any site where things are being visibly worked on. They are irresistibly drawn

FIGURE 18.
Two boys sleeping on the sidewalk, ca. 1930–1950. Photo by Monty La Montaine. ICHi-67115. Courtesy of Chicago History Museum.

FIGURE 19.
Children in Maxwell Street area, ca. 1906. Photo by Charles R. Clark. ICHi-31666. Courtesy of Chicago History Museum.

by the detritus generated by building, gardening, housework, tailoring, or carpentry. In waste products they recognize the face that the world of things turns directly and solely to them. In using these things, they do not so much imitate the works of adults as bring together, in the artifact produced in play, materials of widely differing kinds in a new, intuitive relationship.
WALTER BENJAMIN[130]

Children, like waste and odor, signify spatial disorder on Maxwell Street. They mingle with the profusion of things that nestle up next to each other in other images.

Nathan Lerner/Art Photography

Other than Vivien Maier, perhaps the most notable photographer of Maxwell Street was Nathan Lerner. Lerner was born in 1913 in Chicago and studied at the New Bauhaus (soon renamed the Chicago School of Design), where he was taught by László Moholy-Nagy. Lerner went on to be head of the school's programs in photography and in production design, and in 1946 became director of the school before founding his own company, Lerner Design Associates. He lived within two blocks of Maxwell Street, where he was born and brought up by Ukrainian-Jewish parents who had immigrated in 1905. His father, Louis, was a tailor and his mother, Ida, a painter. Lerner trained as a painter, first at the National Academy of Art and then at the School of the Art Institute of Chicago. He spent almost his whole life in Chicago. His photographs of Maxwell Street, over a hundred of them, were taken between 1935 and 1940, during the Great Depression and before his move to a more abstract style in keeping with the Bauhaus philosophy.

Lerner's personal geography situates him differently from the documentary photographers who were sent into places like Maxwell Street from elsewhere. Lerner was from the area, not just visiting the street and subjecting it to a gaze from outside. He took these photos to practice his art and improve his skills, but had no specific place in which he sought to display them. Indeed, he paid little attention to the images during his long career as an artist and photographer until the 1970s, when he rediscovered them. It was only then that the images began to form part of exhibitions of Lerner's wider work.[131]

Lerner's images are not easily categorized as "documentary," "street," or "art" photography, as they exhibit characteristics of all three. His photographs convey something of the multidimensional character of the marketplace, focusing alternately on the market's "things" and the people who performed the market. The combination of things and the practice of commerce provided an ideal site for an examination of the wider world of Depression-era America.

> Everything, no matter how insignificant and worn, has value—watch gears, buttons, ragged shoes, devastated furniture—and it is the subject of frantic commerce. Maxwell Street is a noisy and populous city to itself, but it was especially so in the Depression, where there was an endless succession of colorful incident, offering, side-by-side, moments of pathos, merriment, the bizarre and the ridiculous, as well as a remarkable example of the human capacity to make do, to get by, to use everything at hand.
>
> **STEPHEN F. PROKOPOFF**[132]

Lerner's images of Maxwell Street spend very little time on context. There are few shots that seek to establish setting, no "landscape" views that attempt to capture the place as whole. Lerner is situated in and on the street. His im-

FIGURE 20.
Final Tribute, *crowd at Maxwell Street, ca. 1935–1940. Photo by Nathan Lerner. ICHi-67124. Courtesy of Chicago History Museum.*

ages focus on details, on faces, on combinations of objects or patterns of light and dark.

While Lerner's work of this period has been described as documentary, his intentions were not journalistic. He does not follow one person through a day's work or concern himself with the presentation of a structured inventory of life on the street. "I knew a lot of people who were completely involved in social-documentary projects, but many of them were really ideologues. I never was. While I had the same sympathies they never crystallized formally, in a Marxist sense."[133]

Composition was clearly at the center of Lerner's creative practice. Light and dark patterns are every bit as central to his interest as the bizarre and macabre juxtapositions the market presented. Lerner's interest in form, texture, and pattern links his more object-focused images to ones that might be thought of as belonging to the documentary tradition. A sea of people in hats and caps, taken from slightly above, seems unconcerned with the lives or actions of the people and more like an image of a pile of shoes or duck eggs.

The intensity of collision that happened at Maxwell Street resulted in images by Lerner and others that straddle the realist documentary traditions of the Farm Security Administration photographers of the 1930s, the provocations of the surrealists, and the abstract pattern-based images of the Bauhaus.

Some of Lerner's images could be mistaken for photographs by Dorothea Lange or Walker Evans. In one image we see a boy with a handkerchief per-

FIGURE 21.
Duck Eggs, *1937. Photo by Nathan Lerner. ICHi-67215. Courtesy of Chicago History Museum.*

forming a trick while the audience looks toward the ground. In another we see the backs and hats of a crowd of men at a rally.

While the subject matter changes from one image to the next, similar compositional forces remain at play: the high contrast, the hats, the backs of people gathered together. And there is always something going on that we can't quite make out: a magic trick, a blues song, an auction or argument. Geoff Dyer has commented on Lange's use of backs and hats to express something of life in the Great Depression.[134] The hat was once a mainstay of menswear and a sign of one's class and standing. As the Depression ground on hats became symbols of resilience. Hats caught Lerner's attention too. Perhaps because they symbolized something about black culture in the 1930s, perhaps because they referred to the status of the street as a place for the purchase of affordable but dapper fashion. As in Lange's images, there is also the play of white highlights against the darkness of the bodies.

FIGURE 22.
People shopping for shoes at shop on Maxwell Street, 1941. Photo by Russell Lee. LC-USF33-012984-M4. Library of Congress. Prints and Photographs Division, FSA/OWI Collection.

There is no evidence that a young Nathan Lerner had any knowledge of the FSA photographers, who were still working when he took his photographs. In 1941 one Farm Security Administration photographer, Russell Lee, would make it to Maxwell Street and take several unremarkable images, including one of people buying shoes (fig. 22).

When Lerner shot his Maxwell Street images, he was in his early to mid-twenties and about to attend the New Bauhaus. The images would not emerge in developed form until the 1970s, long after Lerner had become famous for his later, more abstract studies of light, shade, and pattern. While Russell Lee was taking images to document the United States of the Great Depression for the government, Lerner was looking for other things.

Mannequins

It is almost impossible, for me at least, to see the photograph in figure 23 without simultaneously seeing Dorothea Lange's iconic Migrant Mother, photographed a year earlier, or indeed, numerous other images of bedraggled children that have appeared though the history of documentary photography. But then there is the mask in the right foreground. That does not seem like Lange at all. It feeds into other images Lerner took at Maxwell Street in which people are notably absent or firmly pushed to the background. Masks and mannequins are a frequent theme.

The form of mannequins behind a grating (fig. 25) stands in contrast to the

FIGURE 23.
Young girl holding a child in the Maxwell Street area, 1936. Photo by Nathan Lerner. ICHi-35047. Courtesy of Chicago History Museum.

FIGURE 24.
Head, 1936. *Photo by Nathan Lerner. ICHi-35049. Courtesy of Chicago History Museum.*

weathered bodies of pedestrians outside, a simulacrum of fashion and elegance juxtaposed to the worn and bedraggled bodies of documentary photography. Another Lerner image, *Head, 1936*, shows an open-mouthed head, perhaps a display head used for hats. In the background are both the head of a man and a human skull.

Mannequins have been almost a constant in the history of street photography. The French photographer Eugéne Atget is known for his pictures of things to which no one had previously paid much attention. Mannequins in shop fronts were a favorite subject, and it is not surprising that he attracted the admiration of the surrealists and, particularly, Man Ray.

The surrealists looked for the surprising in everyday Paris. They were fascinated by the slightly out-of-date, the redundant, and, most of all, the marginal.

> As the photographers wandered around Paris like twentieth-century flâneurs, they created a kaleidoscopic vision of Paris through their image of ordinary streets, architecture, store windows with mannequins, flea markets, cafés, dance halls, and anonymous passerby clochards (vagrants), dancers, singers, café society, and prostitutes and criminals of the demi-monde.
> **THERESE LICHTENSTEIN**[135]

FIGURE 25.
Mannequin busts through window, ca. 1935–1940. Photo by Nathan Lerner. ICHi-35044. Courtesy of Chicago History Museum.

In Walter Benjamin's *Arcades Project* the convolute, or file, for mannequins and other kinds of dolls appears immediately after the one labeled photography. Included in this is an extended quote from J. K. Huysmans, which lingers, with the masculine gaze often pinned on the surrealists, on the erotic and nightmarish qualities of mannequins.

> In a shop in the Rue Legendre . . . a whole series of female busts, without heads or legs, with curtain hooks in place of arms and percaline skin of arbitrary hue—bean brown, glaring pink, hard black—are lined up like a row of onions, impaled on rods, or set out on tables. . . . How superior to the dreary statues of Venus they are—these dress-makers' mannequins, with their life-like comportment; how much more provocative these padded busts, which exposed there, bring on a train of reveries.
>
> **J. K. HUYSMANS**[136]

The surrealists had fixed on the mannequin as a key symbol in their artistic armory. A year after Lerner was photographing masks and mannequins on Maxwell Street the surrealists set up their Rue de Mannequins in the Surrealist Exhibition of 1938 in Paris.

Visitors to the show, masterminded by Marcel Duchamp, encountered sixteen mannequins designed and dressed by Dali, Miró, Max Ernst, and others.

FIGURE 26.
Boulevard de Strasbourg, Corsets, Paris, *1912. Photo by Eugene Atget. Gilman Collection, Purchase, Ann Tenenbaum and Thomas H. Lee Gift, 2005, Metropolitan Museum of Art.*

Their bodies were decorated in erotic and macabre ways—a bird cage over a head, flowers in place of pubic hair.

Lerner seems to move between the dual poles of meaning set up by Allan Sekula, between art photography and documentary photography. "The oppositions between these two poles are as follows: photographer as seer vs. photographer as witness, photography as expression vs. photography as reportage, theories of imagination (and inner truth) vs. theories of empirical truth, affective value vs. informative value, and finally, metaphorical signification vs. metonymic signification."[137]

Both documentary and art photography share what Susan Sontag has called the "urge to appropriate an alien reality."[138] This alien reality, in the early years of documentary photography, was the reality of humans who were alien to

FIGURE 27.
Heads. Photo by author.

the middle-class photographers and social scientists who pointed their cameras at the urban or anthropological other. This was what linked the poor of London or New York to the tribes of Africa or Native Americans. But there is another alien reality in the world of objects that we can see in the work of Lerner and others whose images revolved around a simultaneous fixation on art and reality.

I too was open to the charm of mannequin heads on a visit to the market.

Things/ Objects

The world of found objects, Sontag suggests, is a "surreal country" where "our junk has become art."[139] This is a world in which photographers become collectors of the discarded. The project of photography and the project of surrealism thus share, in Sontag's words, the "mandate to adopt an uncompromisingly egalitarian attitude toward subject matter," including "an inveterate fondness for trash, eyesores, rejects, peeling surfaces, odd stuff, kitsch."[140] In Jane Bennett's terms, the photographers are open to the enchantment of objects.[141]

> On a sunny Tuesday morning on 4 June in the grate over the storm drain to the Chesapeake Bay in front of Sam's Bagels on Cold Spring Lane in Baltimore, there was:
>
> one large men's black plastic work glove
> one dense mat of oak pollen
> one unblemished dead rat

FIGURE 28.
Stall selling miscellaneous items at the Maxwell Street Market, September 1959. Photo by Clarence W. Hines. ICHi-67111. Courtesy of Chicago History Museum.

one white plastic bottle cap
one smooth stick of wood

Glove, pollen, rat, cap, stick. As I encountered these items, they shimmied back and forth between debris and thing—between, on the one hand, stuff to ignore, except insofar as it betokened human activity (the workman's efforts, the litterer's loss, the rat-poisoner's success), and, on the other hand, stuff that commanded attention in its own right, as existents in excess of their association with human meanings, habits, or projects. In the second moment, stuff exhibited its thing-power: it issued a call, even if I did not quite understand what it was saying.

JANE BENNETT[142]

FIGURE 29.
Display of wall hangings, 1972. Photo by James Newberry. ICHi-35016. Courtesy of Chicago History Museum.

Bennett describes how she was repelled, dismayed, and entranced by the "impossible singularity" of the things she encountered. The ability of these things to produce affect was heightened by the fact that they appeared together, juxtaposed: "When the materiality of the glove, the rat, the pollen, the bottle cap, and the stick started to shimmer and spark, it was in part because of the contingent tableau that they formed with each other, with the street, with the weather that morning, with me."[143] Bennett ascribes some of the power of her interaction with this collection of things to her own openness to the possibility of "thing-power." Garbage, or waste, is a collection of "things" that have been discarded, that have left the cycle of consumption that ascribed value. It is precisely in this garbage that Bennett finds the power of things. "Look at the power of these things to produce affect," she is saying. And "Look at my openness to things."

Photographers and others who visited Maxwell Street Market were open to the power of things. They often appear overpowered by things. The market is, perhaps, a specially privileged site for the exercise of, and receptiveness to, "thing power."

Photographers and surrealists alike found beauty in places others associated with ugliness and trash. Often these were places where odd and unexpected juxtapositions might occur—places of collision, such as the flea market or secondhand store. Here matter became vibrant.

To Lerner and others, the profusion of objects and their unlikely juxtapositions proved a rich source of images that made the everyday life at the bottom end of retail geography marvelous.

Lerner would refer to the subject of some of his photographs as a form of "natural surrealism." His images are layered with symbolic significance while at the same time being records of actuality. Just as the market mixed the juxtaposition of things with the lively performance of place, Lerner's images combine things and people in such a way that the things are endowed with vitality. Lerner followed a surrealist strategy of enchanting what to many was the waste space of the city. "It is in the actuality of the everyday when passing a second-hand shop, for instance, where umbrellas and sewing-machines find themselves collaged together on a dissecting table. Surrealism is about an effort, an energy, to find the marvelous in the everyday, to recognize the everyday as a dynamic montage of elements, to make it strange so that its strangeness can be recognized."[144]

The flea market has been a steady source of images for street photographers. Flea markets and mannequins are also an almost constant presence in the representational strategies of the surrealists. In André Breton's surrealist novel *Nadja*, for instance, Breton finds delight in the nearby Saint-Ouen flea market. "I go there often searching for objects that can be found nowhere else," he wrote: "old-fashioned, broken, useless, almost incomprehensible, even perverse."[145] The market and the mannequins also resembled the arcades being explored by Benjamin in the 1920s—the collections of things that inspired him to start collecting and listing the detritus of consumption:

> In the arcades, bolder colours are possible. There are red and green combs.
>
> Preserved in the arcades are types of collar studs for which we no longer know the corresponding collars or shirts.
>
> Should a shoemaker's shop be neighbor to a confectioner's his display of shoelaces will start to resemble licorice.

FIGURE 30.
Eye and Barbed Wire, *1939. Photo by Nathan Lerner, © 2018. Courtesy of the David and Alfred Smart Museum of Art at the University of Chicago, and the Estate of Nathan Lerner/Artists Rights Society.*

One could imagine an ideal shop in an ideal arcade—a shop which brings together all métiers, which is doll clinic and orthopedic institute in one, which sells trumpets and shells, birdseed in fixative pans from a photographer's darkroom, ocarinas as umbrella handles.
WALTER BENJAMIN[146]

Benjamin's affection for the juxtapositions of the arcades is mirrored in Foucault's discussion of incongruity in the list from Borges's Chinese Encyclopedia. The impossibly of the list, you will recall, was partly based on the fact that there was no *site* where its contents could actually be—no *table* where "for an instant, perhaps forever, the umbrella encounters the sewing machine."[147]

While Lerner was clearly interested in the uncanny surrealism of life on Maxwell Street there is no sense that he was deliberately pursuing a surrealist agenda—he was simply selecting elements of what the place presented him with. He explored the marvelous forces that emerged from apparently accidental collisions that happen in a hyperplace—barbed wire and an artificial eye, for instance.

The street also presented Lerner with an abundance of form and pattern, and it was perhaps these that were to influence his future career. Jennifer

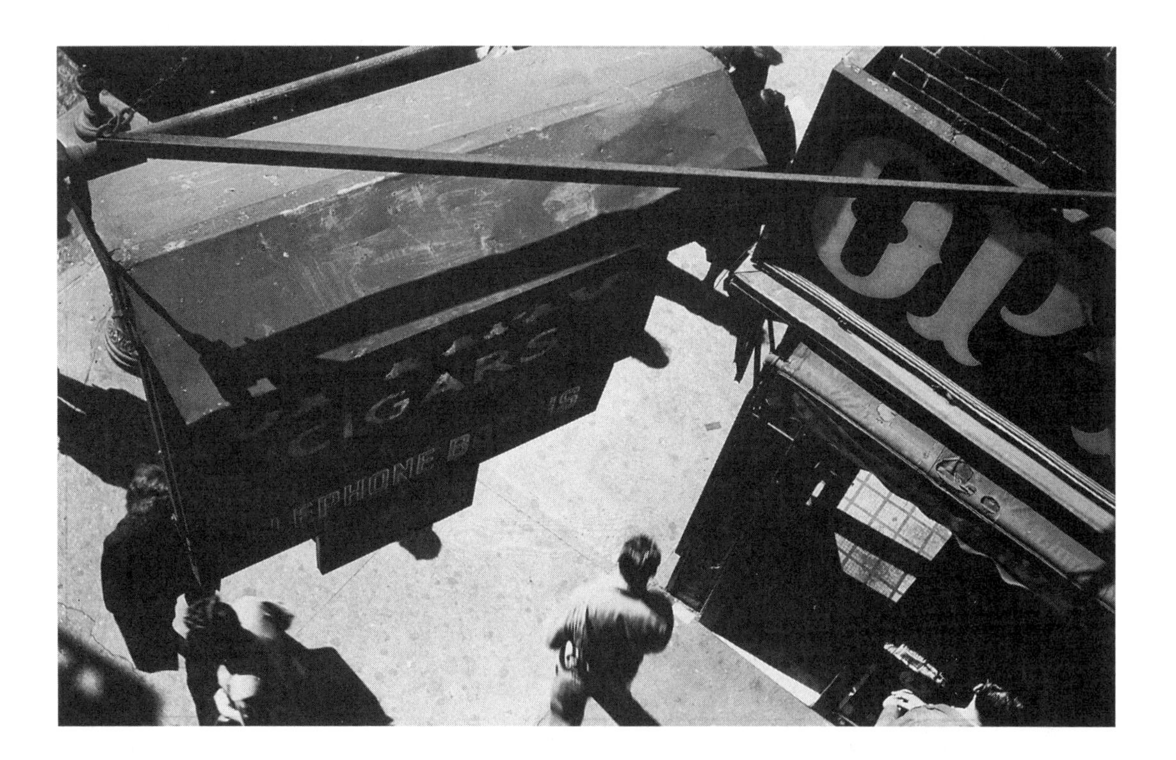

FIGURE 31.
Cigar Store, *1934. Photo by Nathan Lerner. Photo by Nathan Lerner, © 2018. Courtesy of the David and Alfred Smart Museum of Art at the University of Chicago, and the Estate of Nathan Lerner/Artists Rights Society.*

Tucker has commented on how street photographers in the mid-twentieth century began to combine their interest in the actuality of street life with experimentation in "new forms and representational modalities from higher contrast and sharper focus to a strong emphasis on underlying abstract geometries."[148] It is the texture of a pile of eggs that makes Lerner's photograph interesting, just as it is the silvery array of wheels and hubcaps that would later inspire others. Lerner would end up as a student and then a teacher within a Bauhaus tradition that fixated on form and abstraction. He invented a lightbox that allowed him to experiment with geometries of light and dark. While these experiments may appear to be a long way from Maxwell Street, it is clear in his photographs of the market that he was interested in light and pattern all along. When Lerner took a photo from an elevated position it was not to survey the landscape but to see a place differently—to contemplate contrast and pattern from a discombobulating viewpoint.

In an early photograph taken from above at a cigar shop on Maxwell Street, extreme contrast highlights the angular geometry of the image. It is this kind of geometry that he later would replicate in experimental photographs with his lightbox. This interest may also tell us why, when confronted with a store selling used shoes, Lerner was more interested in a close-up of the shoes than in the wider view of people buying them, as in the photo by the FSA photographer Russell Lee (fig. 22).

FIGURE 32.
Light Volume, 1937. *Photo by Nathan Lerner. Photo by Nathan Lerner, © 2018. Courtesy of the David and Alfred Smart Museum of Art at the University of Chicago, and the Estate of Nathan Lerner/Artists Rights Society.*

The Gaze on the Street

In the Maxwell Street image file at the Chicago History Museum are a number of photographs of other people taking photographs. In fact, the Maxwell Street Market proved to be a site of image making.

> While often out of the picture, photographers were themselves part of urban spectacle: stopping for views, unpacking equipment, focusing the lens, and attracting attention from passers-by ranging from curiosity to irritation. Commercial street photographers, viewed as a public nuisance in many cities, were widely prohibited legally from working in urban spaces outside portrait studios.
>
> **JENNIFER TUCKER**[149]

The kinds of objects one could encounter at the Maxwell Street Market have exerted a strange power over generations of photographers and other observers who have stopped to record their curious life on the street. They are momentarily detached from functional relations and an economic system that struggles to encode their value or lack of it. They have been impossible to ignore. A stack of hubcaps or a pile of shoes has its own compositional beauty. What are the histories of these banal objects? What cars did the hub-

FIGURE 33.
Street photographer in the Maxwell Street area, ca. 1906. Photo by Charles R. Clark. ICHi-20788. Courtesy of Chicago History Museum.

caps come from? Who wore those shoes? What stories could they tell? These things come together and appear in a flea market in ways that would not happen in a more carefully organized setting. Flea markets are places of collision and juxtaposition between things and people and the dreams, fantasies, and nightmares of the urban. They are open to the different impulses of the social documentary, the surreal, and the abstract pattern language of the Bauhaus. All of these can be seen in the images of Nathan Lerner.

The photographs of Maxwell Street in the files of the Chicago History Museum were made with different audiences in mind and have followed various trajectories. They include images taken by unknown photographers, by journalists, by members of the Fort Dearborn Camera Club, and by professional photographers. Some appeared in newspapers and magazines, some on the white walls of local galleries. Most had no such public existence. Now they are bundled together in files, divided by decades.

> Streets
> Maxwell Street
> Illinois—Chicago—1900–1909

And so on: 1910–1919, 1920–1929, 1930–1939, 1940–1949, 1950–1959 (in six folders), 1960–1969, and finally 1970–1999. One site of representation for the photographs, then, is the archive. You will also find hundreds of images of Maxwell Street on the internet. There are Flickr pages full of this street.

Perhaps the most surprising place to find images of Maxwell Street is in and around Maxwell Street. As I made my last research visit to Maxwell Street,

FIGURE 34.
Street photographer, 700 West Maxwell Street, August 1946. Photo by James D. MacMahon. ICHi34458. Courtesy of Chicago History Museum.

on October 21, 2012, the market on Desplaines Street was celebrating the centennial of Maxwell Street Market's official recognition by the city of Chicago. I spent some time at the corner of Maxwell and Halsted, the epicenter of the old market. The site is marked with bronze statues of Maxwell Street denizens, mock wooden packing crates (also in bronze), and signs recounting something of the history of the spot. It is here that some of the most striking photographs of the market have reappeared. Some have been made into stylized textured images for the sides of the packing crates (fig. 4). Some accompany the text on the heritage signs. In a dazzling display of circuitous representational acts, one of the pictures on a sign is of a large Hispanic man in a sombrero with three cameras hanging around his neck. On his hat are photos, demonstrating the service he can provide—taking pictures of you at the market to prove you were there. Some photographer took a picture of this provider of photographs. Now that picture is on the street. I took a picture.

After spending an hour watching this intersection I wandered over to the market's new location, passing the silver and glass UIC Forum. As I passed I noticed, through the windows, a series of large-format reproductions of many of the images that had struck me in the archives. The fat photographer was there too—looking back at me through the plate glass. When I arrived at the market, on an unusually warm October afternoon, I followed the sound of blues music and found a Maxwell Street blues band in full swing, celebrating the market's centenary. A good crowd was watching, dancing, taking photos. One man was enthusiastically playing air-guitar. Beside the stage was a stand selling T-shirts for the Maxwell Street Foundation. They were trying to save the Gethsemane church from demolition. Alongside the T-shirts was a display of photographs from across the decades of the market. They were handsome, high-end reproductions, including Nathan Lerner's *Head, 1936*, being sold as a set to raise money for the unending work of saving Maxwell Street.

Value/ Markets/ the City

Places are sites of value. Forms of valuing (and devaluing) help to distinguish a rich sense of place from mere location. Writers and photographers, among others, were valuing Maxwell Street.

> The noises of crowing roosters and geese, the cooing of pigeons, the barking of dogs, the twittering of canary birds, the smell of garlic and of cheeses, the aroma of onions, apples, and oranges, and the shouts and curses of sellers and buyers fill the air. Anything can be bought and sold on Maxwell Street. On one stand, piled high, are odd sizes of shoes long out of style; on another are copper kettles for brewing beer; on a third are second-hand pants; and one merchant even sells odd, broken pieces of spectacles, watches, and jewelry, together with pocket knives and household tools salvaged from the collections of junk peddlers. Everything has value on Maxwell Street, but the price is not fixed. It is the fixing of the price around which turns the whole plot of the drama enacted daily at the perpetual bazaar of Maxwell Street.
> **LOUIS WIRTH**[150]

> MEYER LAZER . . . And my father was one of those peddlers, too. Now in walking down Maxwell Street . . . and there are a lot of kosher butcher shops. And, you know, in a kosher butcher shop, they pull the feathers off . . .
>
> IRA BERKOW Oh, poultry. A poultry store.
>
> ML Poultry, yeah. But here's the point. They . . . I'll say Christian or unkosher poultry shops, when they slaughtered the chicken, they used to dip the bird in hot water to get the feathers off. And all those feathers became wet. The Jew was not allowed to do that. I mean, according to the Jewish law, you're not allowed to immerse 'em in hot water. You have to pluck 'em.
>
> And he used to walk down the street and find these garbage cans with a lot of rubbage and feathers in there. He thought they had value. After all, they're used in Europe and used here too, for beds. So he used to go around to the different butcher shops and make 'em a deal. "Now I'm going to give a barrel. Put

the feathers in the barrel instead of throwing them in with the other garbage, and I'll pay you so much a pound for 'em and I'll pick 'em up every week." . . .

IB How much did he pay them?

ML I have no idea, Maybe 2¢ a pound. Maybe 5¢ a pound. But they realized something. They wouldn't care even if they only took in a dollar a week. It meant another dollar profit.

IB A dollar more than they had otherwise.

ML See, otherwise it was just rubbish. And they have to pay to have the garbage taken away, and here a man comes along and pays them to take this rubbish away. And this is how he started the feather business.
MEYER LAZER, INTERVIEWED BY IRA BERKOW[151]

Exploring a marketplace means returning to one of the root meanings of the city—a place where exchange happens. From the writings of Max Weber to the heretical theory of urban origins proposed by Jane Jacobs to contemporary work in the Marxist political economy tradition, the city is a site characterized by, even originating in, the creation of surplus value through trade.[152] Exploring a market as a place, then, means exploring the most urban of urban sites.

Thus, we wish to speak of a "city" only in cases where the local inhabitants satisfy an economically substantial part of their daily wants in the local market, and to an essential extent by products which the local population and the population of the immediate hinterland produced for sale in the market or acquired in other ways. In the meaning employed here the "city" is a market place.
MAX WEBER[153]

To be fair, the "cities first" argument is still heterodox. Most commentators still believe that urban life emerged from the production of an agricultural surplus, which allowed for a division of labor—with some people growing food and others depending on their labor. But even this version has trade at its center.

The processes of valuing and exchange at Maxwell Street were heterogeneous and multiscalar. The Maxwell Street Market was (and is, though not on Maxwell Street) a flea market. It was a place where a significant portion of what was for sale was secondhand. Shoppers came in large numbers expecting to get bargains. At the same time, the stall-holders expected to, in a telling term, "cheat you fair." The process that ensued was bargaining. This was a practice of valuing that led to (in some instances) the continuing biography of objects as they moved from Maxwell Street to the homes of shoppers from across Chicagoland.

> The sellers know how to ask ten times the amount that their wares will eventually sell for, and the buyers know how to offer a twentieth. Everybody who pushes his way through the crowd is a potential customer, everybody except sightseers, and they are spotted immediately by the discerning eyes of the "pullers," who are engaged in perpetual conversation with the shifting mass of human beings that pass continuously between the rows of street stands piled high with wares.
> **LOUIS WIRTH**[154]

Another form of valuing is the way in which Maxwell Street itself has been consistently valued as a destination—a place to go, a stop on a tourist itinerary as well as a place for a bargain. The visits of photographers to the market over a hundred years reveal a kind of bourgeois valuing of the place of the other—a place to slum it for a day. It was a place to find surreal juxtaposition, picturesque poverty, and constant drama. Different but related forms of valuing can be seen in the works of novelists and sociologists and, indeed, in my own fascination with the place.

Many of the arguments that swirled around Maxwell Street during the process of its erasure were arguments about the values of things. Discussion centered on whether certain objects in the Maxwell Street area, and the area itself, deserved to persist or be discarded. The idea of "regimes of value" suggests certain contexts within which things are ascribed value.[155] It performs a critique of the idea of inherent value and at the same time dispenses with the differentiation between commodity value and gift value (as two subsets of exchange value). Things travel through these regimes and in doing so have "careers" or "biographies" and "social lives."[156] In this sense the objects of Maxwell Street, and Maxwell Street itself, are fluid concretizations of relations between the human and the nonhuman worlds—of the way value is ascribed to objects.

Hubcaps

Photography is one way in which hubcaps in Maxwell Street entered regimes of value. But there are other ways. The hubcap appears repeatedly in the words of those who argued for the demolition and relocation of the market. Its banal materiality became a vehicle for a discourse that framed the market as a site of dubious moral order. In the archive of the University of Illinois at Chicago are a series of letters written to the office of Mayor Richard M. Daley, whose father had been mayor in the era of urban renewal, supporting the relocation of the market. One, from a local merchant, asks:

> Where do the goods come from? On more than one occasion we bought my own hubcaps on Maxwell Street (15 minutes after they were stolen off our car). The absence of this can only have a positive effect on the area and Chicago proper.
> **MICHAEL SHEA TO RICHARD M. DALEY, 1993**[157]

The hubcap was linked to much more serious pronouncements of moral dissolution in a letter from the university's head gymnastics coach.

FIGURE 35.
Hub Caps and Wheels, 1990, Maxwell Street.
Courtesy of Ron Gordon.

When I think of Maxwell Street I think of 3 things:
Garbage
Crime
Perversion....

In regard to crime I personally have witnessed drug deals, prostitution, car thefts, and creeps prowling the area daily. I have to buy back my own hubcaps, radio and accessories two or three times a year.

C. J. JOHNSON TO RICHARD M. DALEY, 1993[158]

The story of finding your own hubcaps at Maxwell Street just after they have been stolen is one of the most often-told stories of Maxwell Street. It is told so often that, in most cases, it is unlikely to be true. Who, after all, would recognize their own hubcaps? This is the way a place becomes storied. A story is told over and over until it sticks—until it becomes so much common sense.

Hubcaps clearly took on great symbolic importance on Maxwell Street. To photographers such as Ron Gordon and James Newberry they presented an aesthetic opportunity, a sign of the object-richness of the market. Piled up in profusion they created form and contrast—and beauty. To others, less enamored of the market, they were signs of a broken society, marking Maxwell Street as an amoral place filled with crime and perversion.

In a society where cars are symbols of access and success, of a speeded-up society, their collision with the flea market throws these meanings into a disconcerting new moral economy.

The assignment of value to hubcaps—specifically the association of hubcaps with crime and marginal space—points toward broader spatialized anxieties with a distinctly racial tone. These quotidian objects become a kind of symbolic currency in a battle of narratives. Value was constantly contested at Maxwell Street.

Stradizookys

In the archives of Ira Berkow, author of *Maxwell Street: Survival in a Bazaar*, are transcripts of all the oral histories he collected for his remarkable book—an account of Maxwell Street collated from those who lived and worked there over the years. Some transcripts didn't make it into the final manuscript, for instance, one documenting an interview with Tyner White—an interview that hilariously goes nowhere:

> IRA BERKOW What does the street mean to you? What does Maxwell Street mean to you?
>
> TYNER WHITE What Maxwell Street means to me? Essentially, it's something that the city means. The city means an exchange market. You visit there, and you offer others things you don't need, and you get things from them that they don't need. These are wares. Wares are things which were. And now I don't need it anymore.
>
> **TYNER WHITE, INTERVIEWED BY IRA BERKOW**[159]

In the home of the Roosevelt University economics professor, Steve Balkin, an advocate for the market in its final years, I noticed some curious wooden objects hanging from shelves. These, he told me, are Stradizookys—musical instruments made from scrap bits of wood and other junk. Tyner White made them. The word "Stradizooky" is a portmanteau of Stradivarius, the renowned violin maker, and Suzuki, whose method is used to teach children to play violin. The Stradizooky combines a passion for recycling wood with a quest for racial/ethnic togetherness. One example in Balkin's loft has "Blacks + Jews = Blues" inscribed upon it.

Tyner White graduated with an MFA in creative writing from the University of Iowa. After a flirtation with poetry he dedicated himself to educating people about the wonders of wood and the necessity of creative recycling. Like many before him he gleaned stuff from the Maxwell Street area to work on his inventions.

> He's built a mad hatter's assortment of prototypes: a possibly functional tape dispenser in the shape of a cat, rubber-tipped walking sticks with handles of telephone wire, oversize sculptural chess pieces sporting shiny metal screws

FIGURE 36.
Stradizookys (Stradizukis) in Steve Balkin's house, 2008. Photo by author.

for arms, a deeply discordant toy violin. "Here," he says, offering a box of lumber scraps that he's sanded and beveled. "Take a diamond."
***CHICAGO READER*, 2005**[160]

Tyner White was a central figure in the Maxworks artists collective who inhabited 716 West Maxwell Street until they were forcefully evicted to make way for the University Village development in March 2002. Theirs was the last inhabited building on the old street. Once evicted, White took his gleaning project to the Resource Center, where he founded the "Maxwood Institute of Treeconomics," which sits alongside the Creative Reuse Warehouse—a place where artists can get scrap materials cheaply for the construction of installations and other artworks.

On New Year's Eve at the turn of the millennium, an old Nabisco factory at 720–724 West Maxwell Street had mysteriously gone up in flames. White and the residents of the Maxworks collective witnessed the fire. Arson was suspected.

> "It's like a war and they're trying to exterminate our resources," says White, who has built thousands of bizarre instruments and knickknacks out of "recycled" scrap wood, including the "Stradizooky," a violin-like musical instrument, and his trademark "Toker," a device for smoking marijuana that he claims will help replace the demand for cigarettes. He says he now hopes to "get a moratorium on bulldozing" and to pave Maxwell Street with bricks salvaged from the demolished factory, which, he says, had also contained remnants from the days when the market was predominantly populated by central European Jews.
> ***CHICAGO READER*, 2000**[161]

White's recollection of the building that burned is linked to his use of recycled wood to suggest an alternative kind of valuation of the place of Maxwell Street and the things in and around it. A valuation based on reuse and recycling rather than destruction and demolition. When I visited the relocated Maxwell Street Market at its new site on Desplaines Street, on its hundredth birthday, there was Tyner White, playing away on a Stradizooky as the Maxwell Street Blues Band did its thing.

The Stradizooky, like the hubcap, is valued in particular ways that are connected to the place it is associated with—Maxwell Street. It tells us about Tyner White's valuation of things that others consider to be junk and evidence for the decay of the Maxwell Street area. A piece of wood becomes a "diamond" or a musical instrument. This particular form of valuation is most evident in another of White's appearances in the distributed archive. He turns up in a report on a public meeting held in October 1993 by the City of Chicago Community Development Commission to consider the future of the market. At the meeting, White pointed out that the University of Illinois at Chicago had a terrible recycling record and offered to take on some of the work through his Maxworks Institute (still on Maxwell Street at the time).

> We could convert some of the scrap lumber into workroom shelves and other kinds of things for the physical plant,
>
> And I would like to mention that in our block are several shuttered buildings which the University acquired over the years, in which they have manifested a wish to tear down.
>
> Now the reason is that ten years from now, it would then be possible to install a four or ten or forty-six million research building. . . .
>
> I would recommend that the University consider recycling the warehouse

buildings on Maxwell Street, make them available for use in a joint venture and find out how much this University can contribute to solving the recycling crisis.
TYNER WHITE, 1993[162]

White's advice was ignored.

In the case of hubcaps and the Stradizooky we have seen how things, at a microgeographic scale, enter and leave regimes of value in a particular geographic context—that of Maxwell Street. They are ingredients in the gathering of things that was Maxwell Street.

Outsider Art

Tyner White was not the only Maxwell Street denizen to reconstitute and revalue the scrap and trash of the market. The profusion of things at Maxwell Street was also a resource for the blues musician and self-taught "outsider

FIGURE 37.
Collage poster by Frank "Little Sonny" Scott Jr. Courtesy of Steve Balkin.

FIGURE 38.
Frank's Blues-Mobile bike at the new Maxwell Street Market, 2007. Photo courtesy of Steve Balkin.

artist" Frank "Little Sonny" Scott Jr. An invitation to the Heartland Café in June 2008 announced a celebration of Scott's eighty-first birthday as well as the fortieth anniversary of the Bugs Bunny Gallery—a surrealist gallery from the 1960s. It was also an opportunity to "remember Maxwell Street." A note in parentheses declared, "(Frank will bring his amazing art posters)." Scott made endless collage posters out of scraps of newspaper and other found images. He also made objects out of found scraps, including walking sticks and a bicycle.

> Outsider artists, in effect, gather up what is usually considered abject in the Kristevian sense, that which the symbolic order discards, such as "trash." Joyce's litter of the letter, the trash and discards, the waste that society throws away, is picked up and reassembled to make a new Imaginary order.
> **JAN JAGODZINSKI**[163]

Tyner White and Frank Scott Jr. both reassembled the trash of Maxwell Street and made it into art. In so doing they ask questions about our relation to the vibrant material world. They remind us that "trash has always been a product of sorting and that what counts as trash has always depended on who was counting."[164]

Assessments of the value, or worthlessness, of hubcaps or bits of scrap wood were symptomatic of a wider assessment of the value of Maxwell Street Market and the area around it. Hubcaps were being mobilized to argue for the erasure of the market. The process of erasure of place had been kick-started by the construction of the Dan Ryan Expressway in 1957. This had been part of the process sweeping urban America (particularly black and working-class America) known as urban renewal.

Urban Renewal

In the mid-1960s the area around Maxwell Street was designated an Urban Renewal Study Area. Urban renewal was a process that began, in Illinois, with the Blighted Areas Redevelopment and Relocation Acts of 1947 and became better funded and nationalized with the federal Housing Act in 1949, which sought to identify so-called slums, knock them down, and replace them with new-model, efficient housing. For the most part it was a process that occurred in the inner-city areas of medium and large cities. The term "urban renewal" more formally entered the lexicon with the 1954 Housing Act, which included the provision of mortgages backed by the Federal Housing Authority to support private developers in the construction of housing for the market.

> Most cities are engaged in . . . something called urban renewal, which means moving Negroes out. It means Negro removal, that is what it means. The federal government is an accomplice to the fact.
> **JAMES BALDWIN**[165]

At the heart of the urban renewal process in Chicago was an expansion of the city government's rights of eminent domain that allowed it to purchase private property and use it for municipal (or "public") purposes. These rights were bestowed by the state of Illinois. Often the property was sold on to private developers, cheap, after undergoing a "write-down" process. This political-economic process was accompanied by a cultural discourse asserting that blight was being eradicated and the city improved.[166]

In practice, urban renewal meant the removal of housing for low-income, mostly black residents. It was this process, as it was practiced in New York, that led to one of the key texts in urban studies, Jane Jacobs's *Death and Life of Great American Cities*, though she was far from the only critic.[167]

> Suppose that the government decided that jalopies were a menace to public safety and a blight on the beauty of the highways, and therefore took them away from their drivers. Suppose, then, that to replenish the supply of automobiles, it gave these drivers a hundred dollars each to buy a good used car and also some special grants to General Motors, Ford and Chrysler to lower the cost—though not necessarily the price—of Cadillacs, Lincolns and Imperials by a few hundred dollars. Absurd as this may sound, change jalopies for slum housing, and I have described, with only slight poetic license, the first fifteen years of a federal program called urban renewal.
> **HERBERT GANS**[168]

There was already a long history of urban renewal in Chicago before the city set its sights on the Maxwell Street area. Urban renewal law had been used, for instance, to clear land inhabited by poor Italian American families to make way for the current site of the University of Illinois at Chicago in 1965. This institution would later become the prime mover behind the erasure of the Maxwell Street Market and the construction of "University Village" in the late 1990s. Another university, the University of Chicago, was behind the most written-about urban renewal scheme in Chicago, in Hyde Park on the South Side.[169] By the time Maxwell Street was being inspected, the urban renewal process had already been met with considerable criticism and was past its heyday.

In order to be designated a zone for urban renewal, the area around the Maxwell Street Market had to be surveyed. The city's Department of Urban Renewal contracted the Institute of Urban Life at Loyola University to produce the "Diagnostic Survey of Relocation Problems of Non-Residential Establishments, Roosevelt-Halsted Area," which was published in 1965. The point of the survey was to assess the problems that would be faced by nonresident establishments due to relocation. For the most part, this meant the stall-holders and shopkeepers of the Maxwell Street Market.

Maps/ Boundaries

The foundation for the designation of urban renewal zones in Chicago was laid long before the 1960s. The creation of territories for instrumental purposes have been layered through history, with each producing the conditions upon which the next is based.

One of the most influential maps to include the intersection of Maxwell and Halsted was the Home Owners Loan Corporation (HOLC) map of 1939. On this map, created at the behest of the newly formed Federal Housing Administration (FHA), the intersection is surrounded by an area tinted red.

> The FHA had adopted a system of maps that rated neighborhoods according to their perceived stability. On the maps, green areas, rated "A," indicated "in demand" neighborhoods that, as one appraiser put it, lacked "a single foreigner or Negro." These neighborhoods were considered excellent prospects for insurance. Neighborhoods where black people lived were rated "D" and were usually considered ineligible for FHA backing. They were colored in red. Neither the percentage of black people living there nor their social class mattered. Black people were viewed as a contagion. Redlining went beyond FHA-backed loans and spread to the entire mortgage industry, which was already rife with racism, excluding black people from most legitimate means of obtaining a mortgage.
>
> **TA-NEHISI COATES**[170]

The tint on the HOLC map labeled Maxwell Street as a high-risk, type-D neighborhood. By making it impossible to get a mortgage, "redlining" prevented residents from creating value through property. The map was thus an

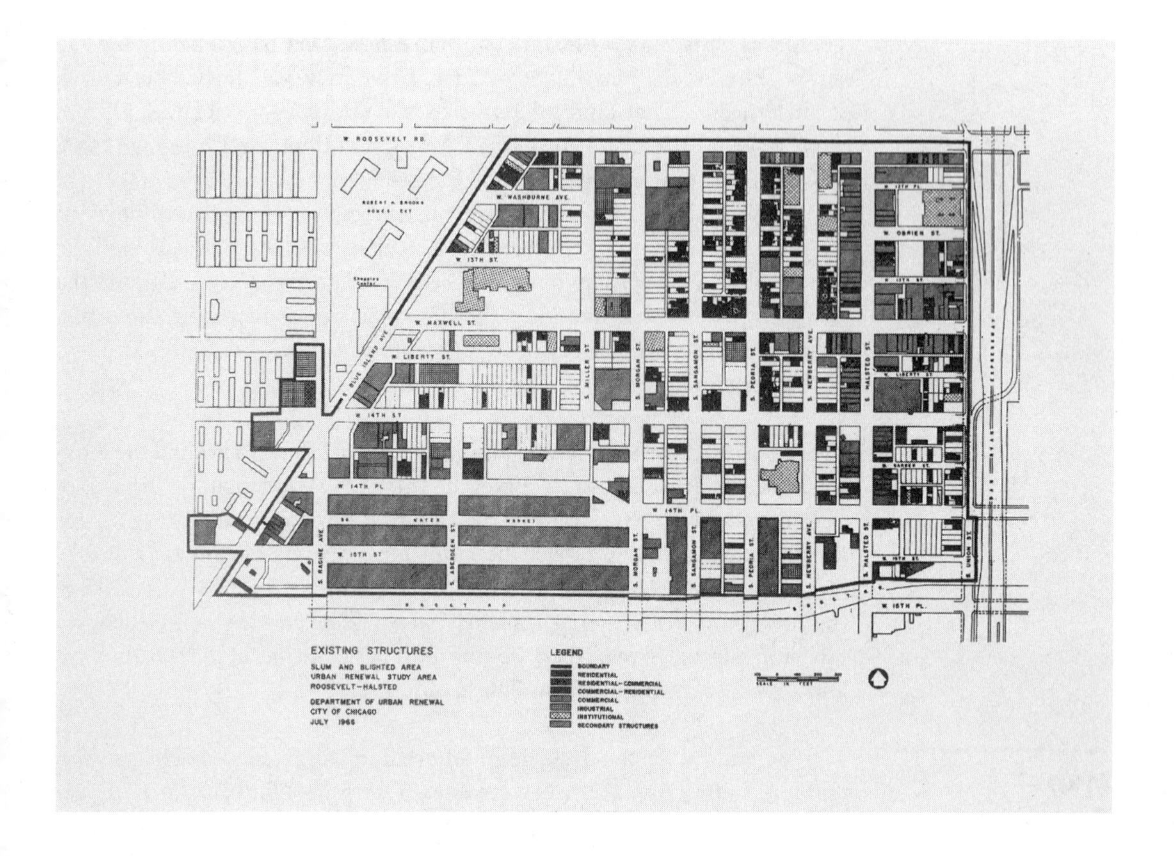

FIGURE 39.
Map of existing structures and textual legal definition, Roosevelt-Halsted Urban Renewal Study Area, Department of Urban Renewal, City of Chicago, July 1966.

instrument in the area's almost inevitable decline—a self-fulfilling prophecy that created the grounds for further deterioration.[171]

The racial work of cartography continued with the creation of "urban renewal zones" in the 1960s.

Maps are both representations of territories and plans for a different world. They suggest a relationship to an actually existing place and the possibility of what that place might become. They are far from passive, and their representational power is itself a kind of doing.[172]

In 1966 another map did the work of defining what *can be* defined as a slum (fig. 39). With an accompanying text specifying borders, it enabled and legitimated a set of surveying practices that resulted in buildings being labeled "old," "dilapidated," "blighted," "overcrowded," and so on. These forms of representation must not be mistaken for description (or at least not only description)—they were doing important work with serious consequences for the future of the area. Both map and text enact precision. The textual account of the area's borders is particularly precise, and this precision is not without poetry. It ends at the point where it starts. It specifies the center lines or particular edges of streets, alleys, an expressway, and a railroad right-of-way as boundaries at which blighted slum ends and something else begins. With few

Beginning at the convergence of the center lines of Roosevelt Road and Blue Island Avenue, thence East along the center line of Roosevelt Road to the center line of Union Street; thence South along the center line of Union Street to the South line of 14th Place; thence East along the South line of 14th Place to the West line of the Dan Ryan Expressway; thence South along said line to the North line of the right-of-way of the B. & O. Chicago Terminal Railroad; thence West along said right-of-way, to the center line of Blue Island Avenue; thence Northeast along the center line of Blue Island Avenue to the center line of 15th Street; thence West along the center line of 15th Street to the center line of Throop Street; thence North along said center line of Throop Street to the South line of Lots 50 to 55 inclusive extended Easterly in Block 9 of William Sampson Subdivision of Blocks 7, 9, 10, 15, and 16 in Sampson & Green's Addition to Chicago; to the center line of the alley next West of Blue Island Avenue; thence North to the extension of the Northeast line of Lot 41, thence Southeasterly along the Northeast line of Lot 41 in Block 9 aforesaid, to the center line of Blue Island Avenue; thence Northeast along said center line of Blue Island Avenue to the South line of Lot 1 (dedicated as a public alley in 1952), extended Easterly of Block 9 aforesaid; thence West along said South line to the East line of Lot 1 (dedicated as a public alley in 1952), of Block 9 aforesaid; thence North along said East line to the center line of 14th Street; thence East along the center line of 14th Street to the East line of Lot 89, extended South, of Block 8 of Sampson & Green's Addition to Chicago being the Northwest 1/4 of Section 20, Township 39 North, Range 14; thence North along the East line of Lot 89 aforesaid to the center line of the alley first North and parallel to 14th Street; thence east along the center line of said alley to the East line of Lot 53 extended south; thence North along the East line of said Lot 53 to the center line of Hastings Street; thence East along said center line to the center line of Racine Avenue; thence South along the center line of Racine Avenue to the center line of 14th Street; thence East to the center line of Blue Island Avenue; thence Northeasterly along the center line of Blue Island Avenue to the place of beginning.

exceptions, it relies on cardinal directions, in keeping with the grid of Chicago streets—themselves the result of cartographic grids that formed the basis for many American city street patterns. Here, this circuit is plotted in thick black lines on the map—carefully placed in the middle of South Blue Island Avenue, West Roosevelt Road, and so on.

> The area covered in this study is known as the Roosevelt-Halsted Area, and is bounded on the north by Roosevelt Road, on the east by the Dan Ryan Expressway, on the South by the Baltimore & Ohio Chicago Terminal Railroad right-of-way, and on the west by Blue Island Avenue. This is one of the oldest portions of the City, having been settled shortly after the Civil War.
>
> **INSTITUTE OF URBAN LIFE**[173]

In their "Diagnostic Survey," researchers from Loyola's Institute of Urban Life construct a history that draws on local community fact books and official records that inform them of the area's incredible growth in the 1860s and 1870s. They show how it became a "slum" through "a haphazard pattern of development, with retail, wholesale and manufacturing establishments mixed in with residential structures."[174] And they quickly arrive at the Maxwell Street Market.

> The latter facility is an open-air market where, on a busy day, more than a thousand individual proprietors arrange their merchandise on temporary stands, at the tailgate of their trucks, or simply on the pavement, in the expectation of selling some of them to passing pedestrians. Since their locations are on the public streets of the area, these entrepreneurs are supposed to obtain a license from the City of Chicago for each day they operate. Many of the operators sell second-hand merchandise, and bargain-hunters come to Maxwell Street from all parts of the Chicago area, and even from locations outside the State. The entrepreneurs who operate in the Maxwell Street Market constitute a location problem which differs from that of the in-store retailer in several aspects. . . .
>
> **INSTITUTE OF URBAN LIFE**[175]

> Activity is intense and exciting on all Sundays when the weather is favorable. Each merchant develops a unique line of merchandise or method of attracting customers. The shopper or browser does not become bored during his tour of the open-air market. Examples of the resourcefulness of the street peddlers abound. One man traps a pigeon and sells it in a cage for $1, while another uses a portable electric megaphone to attract curious passers-by to see and buy his fish "that never have to be fed" (beautiful glass fish for a fishbowl).
>
> **INSTITUTE OF URBAN LIFE**[176]

But the surveyors were not there to enjoy the market, and none of this activity could save the area from being designated as a slum.

> It is clear that the Roosevelt-Halsted area is a slum and blighted area and is eligible for redevelopment under the Urban Renewal Consolidation Act of 1961. That Act defines a slum and blighted area as "any area of land not less . . . than two (2) acres . . . where buildings or improvements, by reason of dilapidation, obsolescence, overcrowding, faulty arrangement or design, lack of ventilation, light and sanitary facilities, excessive land coverage, deleterious land use of layout or any combination of these factors, are detrimental to public safety, health, morals and welfare."
>
> **INSTITUTE OF URBAN LIFE**[177]

The report lays out an array of damning statistics: "97.9% of the 517 structures in the area . . . had one or more of the deficiencies listed in the definition of a slum or blighted area";[178] 66.3 percent of the buildings were deemed dilapidated, 80.7 percent were dilapidated or obsolete; 25 percent of households had resided in their present units for less than one year. These facts, piled one on top of another, led to the conclusion that the area defined by the map was eligible for urban renewal and that it was financially feasible to renew the area with its 584 families and 516 single people.

Numbers. Names. Maps. Narratives. These were all tools in the construction of a particular geography. They were all parts of an assemblage resting on earlier arrangements of similar fragments. And they all aligned to construct particular (im)possibilities for the future. The area around Maxwell Street was made blighted and made into a slum. Actors such as the Institute of Urban Life researchers did not simply record a place. They made a place. And their work rested on earlier work, such as that performed by the HOLC map, which had defined the area as a red zone and therefore ineligible for mortgages. Following the recommendation of the planning department, the Chicago city council declared the area blighted and therefore a slum in August 1966.

Blight

> In other words, we have come to the conclusion speaking in medical terms, that there is a civic cancer which must be cut out by the surgeon's knife—and there is not any figure that describes a slum better. It is exactly like cancer in the human body. It can be cured by radium, if taken in time; but after it has gotten to a certain stage, it infects the body politic, and the only cure for it is to cut with the surgeon's knife. That is what we mean by a Slum Clearance Scheme.
>
> **LAWRENCE VEILLER**[179]

The term "blight" was at the heart of the process of definition being enacted in the various maps and reports that were presented as evidence for renewal status. The maps and textual definitions contribute to an impression that the term is quasi-objective, that blight can be diagnosed and quantified. Yet it remains a vague word, completely lacking in objective specificity, denoting both a cause of a neighborhood decline and a result of that decline.

> The discourse of blight appropriated metaphors from plant pathology (blight as a disease that causes vegetation to discolor, wilt, and eventually die) and medicine (blighted areas were often referred to as "cancers" or "ulcers"). The scientific basis for blight drew attention to the physical bodies inhabiting the city, as well as the unhygienic sanitary conditions those bodies "created."
>
> **RACHEL WEBER**[180]

This use of the term "blight" is central to the work of the Chicago school of sociology, with its predilection for ecological metaphors to describe the city.[181] Insofar as the city was viewed as an ecosystem, blight could be thought of as a natural process. Blight was seen to occur in a "zone in transition" immediately surrounding the central business district of a city. This zone (within which Maxwell Street falls in relation to the central Loop) was also the area of greatest migration and mobility, leading, in Chicago school terminology, to "moral disorder."

There was a clear relationship between race and blight—particularly where African Americans were present in large numbers. Chicago sociologist Ernest Burgess, for instance, noted the "disturbances of metabolism caused by an excessive increase such as those which followed the great influx of southern Negroes."[182] The ecological analogy was taken up enthusiastically by city government and planners. In the early years of urban renewal, reports often included, alongside the presence of large numbers of "Negroes" as evidence for blight, literal medical details such as tuberculosis rates. A moral discourse was tied up with the more scientific senses of the terms used in diagnosing and responding to blight. Blight as a disease in plants is something that spreads, and blight in the city might also spread from one neighborhood to another. It was up to planners to prevent this through a process of renewal.

> The changing terminology used to describe cities set the stage for the implementation of urban renewal. Through the creation and explication of the problem of blight, renewal advocates shifted the terms of the debate. The rights of private property remained sacrosanct, but subject to new limitations. Not all property owners were due the same respect. Those who held onto blighted properties were acting against the public interest because their speculation and inefficient management imperiled city residents and taxed the finances of city government. Furthermore, the refusal of these owners to sell their properties at "reasonable prices" prevented the rationalization of urban real estate and creation of modern cities.
>
> **WENDELL E. PRITCHETT**[183]

The discourse of blight supported and legitimated the process of compulsory purchase, or eminent domain—a process by which a hegemonic belief in the importance of private property is sidestepped in order to find new ways of producing value in the urban environment.

The creativity and lack of objectivity in the use of the term has been made clear by legal scholar Colin Gordon, who recounts how in the St. Louis suburb of Des Peres "local officials declared a thriving shopping mall 'blighted' in 1997 because it 'was too small and had two few anchor stores,' and more specifically, because it didn't have a Nordstrom's."[184]

In Chicago, as elsewhere, the ability to invoke eminent domain relied on the identification of blight based on a list of factors that allowed the city government to designate a lack of value and prescribe renewal. In the three decades following the 1966 report, much of the area it assessed was erased as the University of Illinois at Chicago spread south.

But not without a fight.

The definition of the Roosevelt-Halsted redevelopment area was part of the process by which the market and the landscape around it was erased. It continued the work of the Dan Ryan Expressway, which had carved off the east end of the market in the late 1950s. The Maxwell Street area was characterized by constant change and motion. More was to come.

> The most densely populated area was that east of Halsted Street and south of Harrison Street, already a slum district, especially in the vicinity of Maxwell and Halsted Streets.
>
> ***LOCAL COMMUNITY FACT BOOK, 1960***[185]

The Dan Ryan Expressway represented the first of Mayor Richard J. Daley's major alterations to the spatiality of Chicago. The relocation of the UIC campus was the next. Since 1945 UIC had been located at Navy Pier, north of the downtown Loop on the shore of Lake Michigan. It was always intended as a temporary location. Daley wanted to make sure the campus would stay in or near the Loop as part of his overall strategy for the redevelopment of the downtown area in the face of competition from expanding suburbs. The Illinois legislature's Blighted Areas Redevelopment Act of 1947 and the creation of a Land Clearance Commission allowed Daley to use eminent domain if enough of an area could be considered "blighted." If 60 percent of the residents of an area could be persuaded to move then the other 40 percent could be forcibly relocated. One of the strategies used by the city in order to ensure an area could be declared "blighted" was to let the buildings it owned fall into disrepair. Processes of government disinvestment, then as now, were a key part of the redevelopment toolkit.

> They speak of dwelling units, deficiencies, and relocation. . . . Not mentioned is Tony Lullo's barber shop at Halsted and Taylor where he has been cutting hair for 40 years. Nor is the Taqueria Mexico on Halsted near Roosevelt where you can get tortillas with ranch style eggs swimming in hot sauce. One of these buildings in this slum and blighted area is the two story shingled house of Mrs. Marie Pelletiere. Her kitchen is nearly done in yellow and white, her living room in pink. . . . "Slum?" she says, her voice rising. We've got homes, beautiful homes, here.
>
> ***CHICAGO DAILY NEWS*, 1961**[186]

> First, most Negroes and other minority group members will be swept out of the area, or rigidly confined to their public housing compounds. Exiled to their projects they will be demoralized, and soon the call will go out for more social workers, more inter-group workers, and above all, more police. Secondly, the entire section will resemble the Medical Center, Lake Meadows, and Hyde Park. That is, drab by day, and moderately dangerous by night. The new campus, the new high rises, everything will take on a garrison atmosphere. More and more private police would be needed. Thirdly, the animating principle of the new Near West Side will be segregation. The races will be segregated, the poor will be segregated, education will be segregated, work will be segregated. Fourthly, the Medical Center and the University of Illinois will ask for the clearance of the non-public housing areas south of Roosevelt Road. Fifthly and last, the old inhabitants, their communal and civic life fractured, . . . will give up the ghost and flee.
>
> **MONSIGNOR JOHN J. EGAN, 1961**[187]

In 1956 the largely working-class Italian Harrison-Halsted area (one block north of Maxwell Street) was declared blighted and slated for demolition. The outcome of the "renewal" was to be the UIC Circle campus, named for the "Circle Interchange" between the newly constructed expressways. The process was protested and litigated, eventually reaching the Supreme Court—which ruled in 1962 that residents of the area had no right to be heard as they did not constitute the city as a whole. Only the City of Chicago had legal standing. Protestors succeeded in saving only Jane Addams's Hull-House from demolition. It is now somewhat uncomfortably nestled amid the fortress campus of UIC on Halsted Street.

Once the area was redeveloped, and UIC had moved, Maxwell Street became the barrier to further campus expansion, as it has not been included in the Harrison-Halsted area. The story of the next decades was UIC versus Maxwell Street.

> For some reason Maxwell Street was excluded. It drove us up the wall. When you looked at it, Maxwell Street was the area that should have gone because it was such a slum, but there was never a focus there. There never seemed much of a community rallying around it, either. The people were very poor and disorganized, and, besides, there wasn't much for them to keep.
> **FLORENCE SCALA**[188]

> It is important that development of these uses proceed with full recognition of not only the need for efficient land and structure use, but that all the elements from sign lettering, to plaza pavement, to brick color and texture, involved in urban design, are considered. These need to be utilized in harmonious and restrained manner so that the resident, the shopper in the market, the worker in the industry, and the student in the University recognizes and reacts positively to the renewed environment.
> **CHICAGO DEPARTMENT OF URBAN RENEWAL**[189]

Planners/ Planning

Simone de Beauvoir and Willard Motley were Maxwell Street writers, but so were the inspectors and surveyors, directly or indirectly working for the City of Chicago, who wrote a series of reports on the area surrounding the market, and had a more direct effect on the place itself.

1951: West Central Area Report—produced for the Chicago Plan Commission
1958: Development Plan for the Central Area of Chicago—Chicago Department of City Planning
1965: Analysis of the Maxwell Street Market—Chicago Department of Urban Renewal
1965: Diagnostic Survey of Relocation Problems of Non-Residential Establishments, Roosevelt-Halsted Area—Institute of Urban Life, Loyola University, for the Chicago Department of Urban Renewal
1966: Economic and Market Analysis, Near West Side, South of Eisenhower Expressway—Chicago Department of Urban Renewal

1966: Designation of Slum and Blighted Area—Roosevelt-Halsted—Chicago Department of Urban Renewal

Planning reports defined an area, provided maps, and concluded with recommendations. There was a direct relation between the place as reported by the authors and visions of what the future place might look like. Each report cited the reports that preceded it as part of its evidence. Thus, narratives built on narratives in a kind of dialogue between plans and place.

None of these reports defined the place they were reporting on as "Maxwell Street." Rather, the street existed within a series of officially designated areas with different purposes. The market, and later the Roosevelt-Halsted Urban Renewal Area, existed within the Roosevelt-Halsted Shopping Area, as defined in the 1954 census of business. It fell also within the Near West Side community area defined by the Social Science Research Committee of the University of Chicago in the late 1920s. These names enact the definition of particular territories, and each informs the production of the next.

Despite the instrumentality of the writing practice of planners and surveyors, the texts they produced over several decades contained moments of semi-ethnographic observation. These were, after all, people with their own aesthetic predispositions, wandering the streets around Maxwell Street and paying close attention.

> 1. The Market has some aspects of the circus or fair, and shopping there, at least for some of its visitors, is a form of entertainment as well as method of purchasing needed goods.
>
> 2. The Market is conducted in the open air, in an informal, almost disorderly manner. The place where the customer walks in order to observe goods on display is a paved public street, where an occasional automobile challenges the crowd of pedestrians for the right to move. The goods offered for sale are displayed on makeshift tables, at the tailgate of trucks or station wagons, and sometimes on newspapers spread on the street. In addition to the stationary merchants, some vendors circulate through the crowds carrying their merchandise, or wheeling it in a handtruck. Many of the vendors shout advertisements for their wares, sometimes using loudspeaker equipment.
>
> On 14th Street a small card table, manned by two men in their mid-twenties, was filled with magic tricks and novelties. These men stated that although they had been operating for only six weeks, they have attended every Saturday and Sunday during this time. Both of these men have other jobs which are their principal sources of income. One of the men is also a magician and on occasion performs for various groups for pay. Certainly the brisk pace of sales at this stand was primarily due to the fact that one of the young men was performing magic tricks to induce customers to stop.
>
> **INSTITUTE OF URBAN LIFE**[190]

Erasing Maxwell Street

The area surrounding Maxwell Street had been an immigrant place since the 1840s. As is often the case with such places, it had also been marked by extreme poverty. The question of value was always key. The sociologists of the Chicago school of sociology would recognize Maxwell Street as firmly located in the "zone in transition"—the space where immigrants inevitably first arrived only to move upward and outward as they prospered. The models of the city produced by the Chicago school had land value and rent as the bedrock of their explanations—explanations mobilized by Louis Wirth in his ethnography of the "ghetto."

> Land values rose from the level of farm land to centrally located urban real-estate levels. The streets and the buildings soon became inadequate, and the neighborhood rapidly deteriorated. Property owners saw no reason for undertaking improvements, for the rent they could squeeze out of their holdings did not warrant costly repairs, especially since their property was located on the very edge of the central business district, and would therefore, within a few years, be more valuable for industrial sites than residential purposes. Meanwhile, however, several generations of immigrants found this area their temporary living quarter.
>
> **LOUIS WIRTH**[191]

Immigrants with little money gravitated toward low-rent areas that were close enough to their likely places of work in the central business district. Maxwell Street was within walking distance of the Loop.

From its inception to the day of its erasure, the Maxwell Street Market was the poor part of town. The landscape was neglected and allowed to fall into disrepair. The origins of the people who lived in the area changed from Irish to Jewish and Eastern European to African American and Latino. They were always poor and marked as, in some way, other.

Before the construction of the UIC Circle campus, city government had shown little interest in improving the conditions of the people living there. Once the university arrived, Maxwell Street became an impediment to further expansion of its campus. It was at this point that the city government and UIC began to see value in the land. What had been a place for the poor to live, in low-rent homes on low-value land, became a possible source of value and a place where the relatively rich might live.

In 1987 the Strategic Planning Committee of the University of Illinois at Chicago produced a document called "A Look to the Future: Strategic Plans for UIC," outlining ways to strengthen the university's research profile by attracting renowned faculty and expanding its facilities.[192] Early the next year the university started to purchase land in the area through small trusts and requested an obliging City of Chicago District Development Commission to declare a moratorium on the sale of city land in the area to other buyers. The city's Department of Economic Development obtained the title to local tax-delinquent properties and began the process of demolishing properties considered "blighted." By the end of 1988 UIC had procured more than forty parcels of land in the Maxwell Street area. In the following years the properties were simply abandoned, preparing the way for future verdicts of dereliction and blight. The landscape architecture firm Johnson, Johnson and Roy was hired to prepare a new campus master plan.

The goal of expanding UIC's facilities was impeded by the presence of Maxwell Street Market. A further report was commissioned to look at the status of the market and its relation to UIC.

> One of the joys city dwellers possess is the vibrant perseverance of creativity and utility found in out-of-place, contradictory places. It is in these clefts broken by history or chance into a stony urban edifice, forgotten by regulation, that good things often recur and survive with an order of their own. Planning attempts to preserve the good of these places while minimizing their ill effects.
> **CHICAGO DEPARTMENT OF PLANNING, 1989**[193]

Among the goals and objectives of the proposed Master Plan for the South Campus Development:

> Use the South Campus development as a catalyst for building a vibrant UIC community in place of a blighted area.
>
> Use Private and non-traditional funding sources to implement the plan.[194]

The UIC East/South campus plan map (fig. 40) is notably different from the urban renewal maps of the area. While the latter are stark and functional the former is full color—and notably green. Trees appear on the plan for the area's future. It is possible to imagine living there and using the space.

Tax Increment Financing

In 1997 the area around Maxwell Street called "Roosevelt-Union" was valued at $3,968,563.

What are the processes that take us from one landscape, widely seen as a landscape of decay and dereliction, to another—a landscape of ordered aesthetics? How do we get from crumbling wood and stone to wrought-iron fences via the global bond market? How do boundary definitions simultaneously construct territory and pull it apart?

Each map, each boundary definition, and each report on the Maxwell Street area enacted territorialization. Each one bound the gathering of place into a particular configuration of materialities, meanings, and practices that could, and could not, exist there. Less obviously, perhaps, they connected Maxwell Street to other sites through the transport of stories and ideas from elsewhere. The definition of the area as a tax increment financing (TIF) district, for example, has its origins in California, where the idea of tax increment financ-

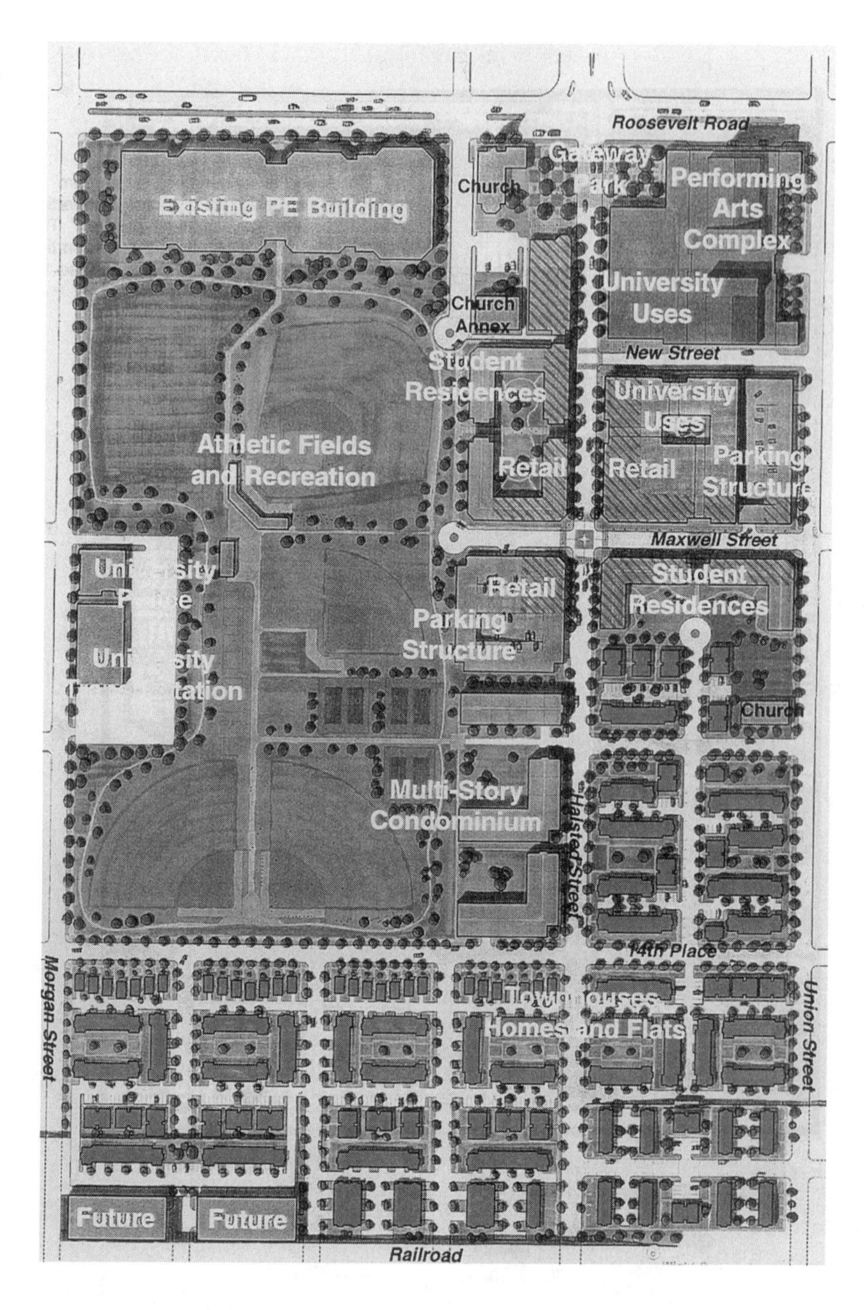

FIGURE 40.
Map of east/south campus plan, 1998. University of Illinois at Chicago.

ing was born in 1951. Such stories and ideas connected Maxwell Street into flows of value.

This is how Maxwell Street was described in the 1980s:

> During the week, the dusty vacant lots are more desolate than ever. Shabbily dressed old men sit silently on crumbling stoops and drink wine in garbage-strewn alleys; the few remaining buildings sag wearily, burned out stairwells

and boarded-up windows telling the perennial urban story of neglect and decay.

***CHICAGO READER*, 1988**[195]

A pamphlet mailed to UIC employees in 2001 advertised the area in Far different terms:

"Yesterday's Heritage. Tomorrow's Treasure"

Chicago's newest, most convenient, most thoughtfully planned neighborhood . . . a great life in the city.

University Village presents traditional Chicago style architecture on tree lined streets

Townhome exteriors feature varying rooflines.

The site plan features neighborhood parks and green space corridors.

"Chicago's next great neighborhood"[196]

It is necessary to connect the materiality of place (by which I mean, in this instance, things like limestone, wood, and wrought-iron fencing) both with meanings and narratives of place (such as "blight") and with practices that happen in place (such as market selling and demolition). These three aspects of place need to be understood collectively in order to fully appreciate the processes by which places change or endure.

The dereliction of the 1980s was *necessary* for the existence of University Village in the present day. Not just because the landscape needed to be derelict in order to be demolished but because the dereliction could be made valuable. Acts of territorialization and deterritorialization were central to this process. These form part of what David Harvey, following Marx and Schumpeter, refers to as "creative destruction."[197] "Valuable assets are thrown out of use and lose their value. They lie fallow and dormant until capitalists possessed of liquidity choose to seize upon them and breathe new life into them."[198]

The truth of the matter, as Marx sees, is that everything that bourgeois society builds is built to be torn down. "All that is solid"—from the clothes on our backs to the looms and mills that weave them, to the men and women who work the machines, to the houses and neighborhoods the workers live in, to the firms and corporations that exploit the workers, to the towns and cities and whole regions and even nations that embrace them all—all these are made to be broken tomorrow, smashed or shredded or pulverized or dissolved, so they can be recycled or replaced next week, and the whole process can go on again and again, hopefully forever, in ever more profitable forms. The pathos of all bourgeois monuments is that their material strength and solidity actually count for nothing and carry no weight at all, that they are blown away like frail reeds by the very forces of capitalist development that they celebrate. Even the most beautiful and impressive bourgeois buildings and public works are disposable, capitalized for fast depreciation and planned

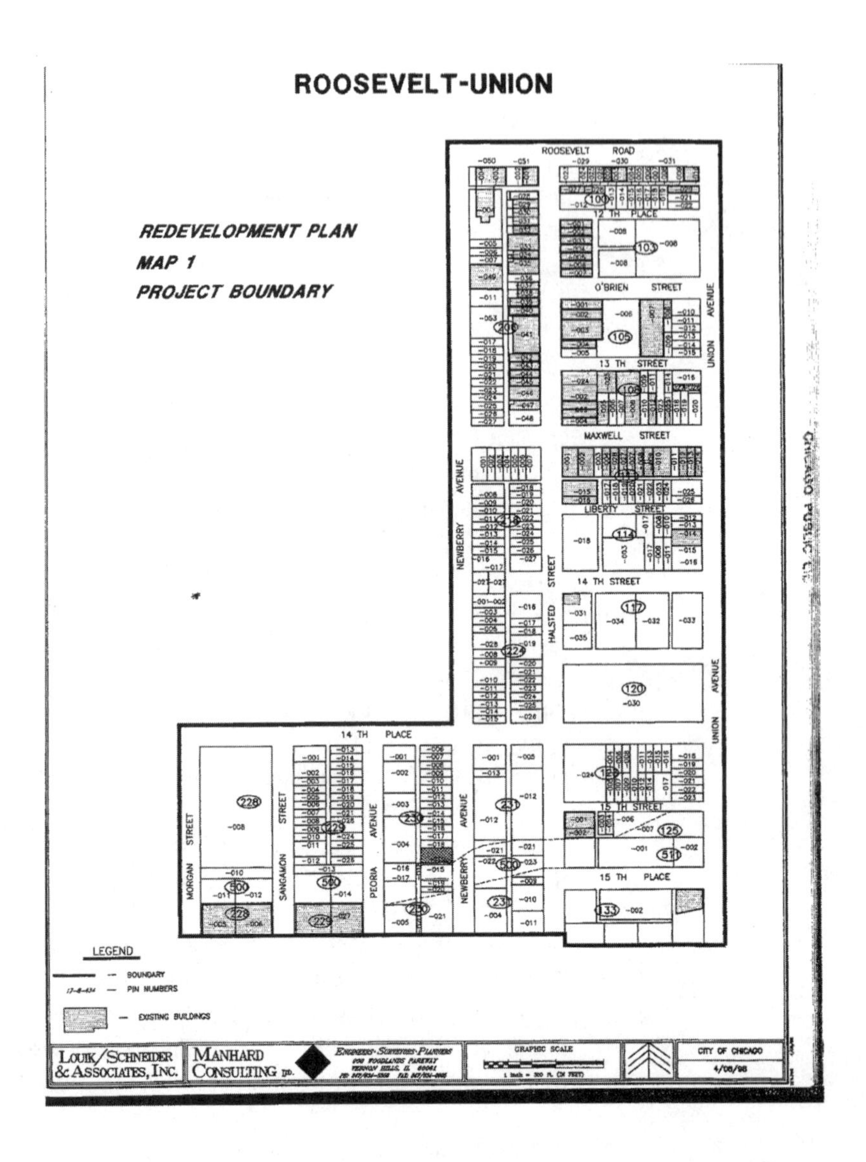

FIGURE 41.
Map of Roosevelt-Union TIF district and corresponding textual definition.

> to be obsolete, closer in their social functions to tents and encampments than to "Egyptian pyramids, Roman aqueducts, Gothic cathedrals.
>
> **MARSHALL BERMAN**[199]

In Chicago, as elsewhere, the city was able to purchase land and prescribe renewal only after diagnosing blight and thus designating the parcels' lack of value. In the following thirty years much of the Maxwell Street area was erased as the University of Illinois at Chicago spread south. The market remained, in a diminished form, until 1994.

The fluid, restless nature of capital constantly comes up against the friction of fixed capital—the relative intransigence of bricks and mortar or, to bring the metaphor up to date, glass and steel.[200] One way to navigate this problem and

IAT PART OF THE EAST HALF OF THE NORTHEAST QUARTER OF SECTION 20 TOWNSHIP NORTH RANGE 14 EAST OF THE THIRD PRINCIPAL MERIDIAN AND THE WEST HALF OF IE NORTHWEST QUARTER OF SECTION 21 TOWNSHIP 39 NORTH RANGE 14 EAST OF THE IIRD PRINCIPAL MERIDIAN, DESCRIBED AS BEGINNING AT THE INTERSECTION OF THE)RTHERLY EXTENSION OF THE WESTERLY RIGHT-OF-WAY LINE OF NEWBERRY VENUE AND THE CENTERLINE OF ROOSEVELT ROAD; THENCE EASTERLY ALONG SAID :NTERLINE OF ROOSEVELT ROAD TO THE NORTHERLY EXTENSION OF THE EASTERLY GHT-OF-WAY LINE OF UNION AVENUE; THENCE SOUTHERLY ALONG SAID NORTHERLY (TENSION AND EASTERLY RIGHT-OF-WAY LINE TO THE EASTERLY EXTENSION OF THE)UTHERLY LINES OF LOTS 14,15 AND 16 IN CANAL TRUSTEE'S NEW SUBDIVISION IN IE NORTHWEST QUARTER OF SECTION 21 TOWNSHIP 39 NORTH RANGE 14 EAST OF THE IIRD PRINCIPAL MERIDIAN IN COOK COUNTY, ILLINOIS, RECORDED MAY 17, 1852; IENCE WESTERLY ALONG SAID SOUTHERLY LINE, SAID LINE ALSO BEING THE ORTHERLY RIGHT-OF-WAY LINE OF DEPOT STREET TO THE EASTERLY RIGHT-OF-WAY INE OF HALSTED AVENUE; THENCE NORTHERLY ALONG SAID EASTERLY RIGHT-OF-'AY LINE TO THE EASTERLY EXTENSION OF THE SOUTHERLY LINES OF LOTS 7 AND 26 I BLOCK 30 IN BARRON'S SUBDIVISION OF BRAND'S ADDITION TO CHICAGO, BEING A UBDIVISION IN THE EAST HALF OF THE NORTHEAST QUARTER OF SECTION 20 OWNSHIP 39 NORTH RANGE 14 EAST OF THE THIRD PRINCIPAL MERIDIAN RECORDED UNE 10,1861; THENCE WESTERLY ALONG THE SOUTHERLY LINE OF LOTS 7 AND 26 IN LOCK 30 AND ALONG THE SOUTHERLY LINES OF LOTS 7 AND 26 IN BLOCK 29 AND HEIR EASTERLY AND WESTERLY EXTENSIONS IN SAID BARRON'S SUBDIVISION OF RAND'S ADDITION TO CHICAGO TO THE SOUTHWEST CORNER OF LOT 7 IN SAID BLOCK 9; THENCE WESTERLY TO A POINT ON THE EAST LINE OF BLOCK 28 IN BRAND'S DDITION TO CHICAGO, BEING A SUBDIVISION OF THE EAST HALF OF THE NORTHEAST QUARTER OF SAID SECTION 20, SAID POINT BEING 164.41 FEET SOUTH OF THE NORTHEAST CORNER OF SAID BLOCK 28; THENCE WESTERLY TO A POINT ON THE WEST LINE OF SAID BLOCK 28, SAID POINT BEING 164.37 FEET SOUTH OF THE NORTHWEST CORNER OF SAID BLOCK 28; THENCE WESTERLY TO THE SOUTHEAST CORNER OF LOT 26 IN BLOCK 27 IN SAID BARRON'S SUBDIVISION; THENCE WESTERLY ALONG THE SOUTHERLY LINE OF LOTS 7 AND 26 IN SAID BLOCK 27 AND THE WESTERLY EXTENSION THEREOF TO THE WESTERLY RIGHT-OF-WAY LINE OF MORGAN STREET; THENCE NORTHERLY ALONG SAID WESTERLY LINE TO THE WESTERLY EXTENSION OF THE NORTHERLY RIGHT-OF-WAY LINE OF 14TH PLACE; THENCE EASTERLY ALONG SAID EXTENSION AND SAID NORTHERLY LINE OF 14TH PLACE TO SAID WESTERLY RIGHT-OF-WAY LINE OF NEWBERRY AVENUE; THENCE NORTHERLY ALONG SAID LINE TO THE POINT OF BEGINNING, ALL IN COOK COUNTY, ILLINOIS.

reduce the friction is to create financial instruments that bundle and abstract the idea of landscape or place value and make it transferable.

> Prior investments create path dependencies that, because of the difficulties inherent in modifying physical structures, constrain future investments. The temporal horizons of investors, developers, and residents rarely coincide. The very materiality of the built environment sets off struggles between use and exchange values, between those with emotional attachments to place and those without such attachments.
>
> **RACHEL WEBER**[201]

In 1999 the City of Chicago decided to designate the area around Maxwell Street as the Roosevelt-Union Redevelopment Project Area under the Tax Increment Finance Program. The ideology and practice of urban renewal had fallen out of favor following the popular critiques of Jane Jacobs, the Black Power movement, and others. By the 1990s the far subtler TIF mechanism was in place to define areas of the city as in need of improvement.

In Chicago, the TIF process has been at the heart of city-supported urban development for several decades and is still going strong. TIF works by allocating as yet unrealized increases in property taxes from an area approved as a TIF zone to pay for "improvements" in that area. The idea, as I've said, was not native to Chicago but had been pioneered in California as early as 1951. Forty-eight states have since embraced it as a means of financing improvements in infrastructure and economic development. An Illinois state law permitting the practice had been enacted in 1977, but the first Chicago TIF district—the Central Loop—did not come into existence until 1984. Since then, the City of Chicago has used tax increment financing in more than a hundred areas of the city and has been its most enthusiastic advocate.[202]

The story of TIF is the story of how a designation of lack of value (blight, slum, ghetto, disorder) becomes the basis for a process of securitization and financialization that produces new and different value. It is a process worthy of a Maxwell Street magician.

When the designation of "blighted area" is made, a tax baseline is determined based on assessed property values at that moment. This assessment is then frozen and, for the following 23 years, only the revenue associated with this tax base will be available for citywide school districts, roads, parks, or other general civic amenities. Any additional tax revenues collected during this period—that is, any *tax increment* resulting from new development—will be used to finance further development within the TIF district, paying for district-specific public works or subsidies to private developers. Developers can thus begin work based on future (as yet unrealized) tax revenues. This process favors big developments on large parcels of land where large increases in tax revenue can be quickly realized.

As with urban renewal programs, the TIF process uses eminent domain to purchase land. In contrast to urban renewal, the proceeds almost always flow directly to private property developers, with little transparency or public oversight. Many argue that this diverts money from public bodies that would otherwise benefit from increasing tax revenues.

Once a TIF district is dissolved, at the end of its twenty-three-year period, the city should benefit from the increased tax base resulting from the development process.

Obviously, there is a catch here: the increased tax revenues are not available in advance. The local government therefore has to invent a financial instrument to provide funds up front. To do this it issues bonds with future tax revenues as security. The bonds are sold through negotiated sales to a variety of investors including global pension funds. "In this way, cities obtain capital by turning the rights to their own heterogeneous property tax base into standardized tradable assets—often without the knowledge of the individual property owners paying the tax bills."[203]

Risk/ Capitalization/ Securitization

The TIF process puts the landscape into a vast, complicated, and fragile network of distributed risk. The city government cannot be certain of the increase in tax revenues that serve as collateral for the investment, as the main drivers of property value are situated well beyond the TIF zone or even the city as a whole. Part of this process has been outlined for another TIF area in Chicago, Cabrini-Green:

> The City developed a complex financing scheme to secure the TIF bonds because of the political risks inherent in the project.... It paid a large fee to the Bank of Canada to provide bond guarantees and arranged for Nations Bank to engage in an interest-rate swap that gave Nations Bank the right to invest the tax increments. Only with these costly guarantees to nullify the political risk would the insurance funds purchase the TIF bonds.
>
> **RACHEL WEBER**[204]

This is part of what Andrew Leyshon and Nigel Thrift have called the "capitalization of almost everything."

> We are in a period in which the process of securitization, which has driven so much of international finance since the 1980s, is engaged in a fresh round of tracing value to its source—or rather sources—since what we can see now is an impulse to identify almost everything that might provide a stable source of income, on which more speculation might be built, being brought into play.
>
> **ANDREW LEYSHON AND NIGEL THRIFT**[205]

This reconfiguration of securitization brings speculation into contact with the economics of everyday life as it happens in situ. It is everyday life that is being securitized.

Securitization refers to the way in which a loan is secured. The most obvious example is the way in which a mortgage is secured by the value of the property being bought. If you cannot pay your mortgage, then the bank has your house to recoup its losses. Leyshon and Thrift argue that this process is increasingly turning to previously moribund elements of the physical landscape to find new forms of security that will enable the wilder processes of speculation that have been identified as the frontier of contemporary financial capitalism. These moribund elements, now being used for security, include infrastructure such as roads, parks, and water systems—previously thought of as the least exciting components of the financial landscape.

> Whether it is the micro-geography of leasehold properties, the regional provider geography of public facilities, or the national geographies of the poor and those at risk, what is being sought out are spaces that can be constructed as assets for international finance.
>
> **ANDREW LEYSHON AND NIGEL THRIFT**[206]

These previously overlooked elements of our everyday world are being inserted into the financial system through innovative bundling of assets produced using computer software. This makes it possible to sell bonds with new kinds of security, bonds that become part of a global system of pension funds and the like that are seeking diversity in their assets and risks.

The Roosevelt-Union area surrounding Maxwell Street was made a TIF district on May 21, 1999, and will cease to be one on May 21, 2022. At the time of its inception it was one of around seventy TIF districts in Chicago, most of which had been approved in the previous two years. In 1999 around 7.7 percent of all property in Chicago was in TIF zones.[207] As with all TIF zones the Roosevelt-Union district underwent an eligibility study in order to ensure that it counted as "blighted." The study was hired out to a consultant—Louik-Schneider and Associates, Inc.—which submitted an eligibility plan to the city's Community Development Commission. The CDC, which has never turned down an application, was in turn obliged to order a public hearing (and to give notice to property owners in the area, not the wider public, fourteen days before the hearing). The public hearing has no standing, and the city can thus choose to do what it wants following the CDC vote. The process then passes through the city council's Finance Committee and the council as a whole. Following this process the fifty-eight acres of Roosevelt-Union, with a 1997 "equalized assessed" property value of $3,968,563, became a TIF district.

The Roosevelt-Union Redevelopment Plan and Project was published in October 1998. Its stated objective was to "encourage mixed-use development, including new residential, institutional and commercial development within the Area" and to enhance the city's tax base and preserve the values of existing property. The report states that the area indicated in the map and defined in text (as with the urban renewal area thirty-two years earlier) is well suited to mixed use due to the proximity of transport infrastructure, including Chicago Transit Authority bus and train lines and major highways.[208] The area is defined in relation to other areas that it is either part of or overlaps. These include the Near West Side Community Area and the Roosevelt-Halsted Urban Renewal Area. It also mentions the University of Illinois at Chicago and the Maxwell Street Market as local institutions. The area, the report notes, was formally designated a slum and blighted area on August 11, 1966, and is therefore eligible to be a TIF. In addition, it cites a number of other documents and policies that would support the TIF designation, including the University of Illinois Master Plan of 1977 and the 1996 Chicago Zoning Ordinance.

Aesthetics

These general details are interspersed with "design objectives" for the improved area, including "high standards of appearance" and the need to encourage "a variety of streetscape amenities which include such items as sidewalk planters, flower boxes, plazas, variety of tree species and wrought-iron fences where appropriate."[209]

At first glance, such details seem more than a little strange. Given the scale of investment in a TIF district and the context of global finance, details like flower boxes and wrought-iron fences appear very marginal. They are, however, central to the process of redevelopment of which TIF forms a part. Chicago has a long history of mixing aesthetics with urban planning, most famously, perhaps, in the "city beautiful" movement associated with the influential urban planner and architect Daniel Burnham, who insisted on the importance of grand and beautiful buildings to the well-being and morale of the populace.

> Make no little plans. They have no magic to stir men's blood and probably themselves will not be realized. Make big plans; aim high in hope and work, remembering that a noble, logical diagram once recorded will never die, but long after we are gone will be a living thing, asserting itself with ever-growing insistency. Remember that our sons and grandsons are going to do things that would stagger us. Let your watchword be order and your beacon beauty.
> **DANIEL BURNHAM**[210]

In his 1909 plan for Chicago, Burnham sought to reconfigure the city through the construction of parks, boulevards, and grand, beautiful public buildings inspired by Haussmann's renovation of Paris. Burnham believed that the aesthetics of this new city would uplift the masses. It was, as Peter Hall, has described it, "trickle-down urban development."[211] It was also trickle-down aesthetics. The ideology behind it was that the beauty of parks and museums would benefit everybody.

At the end of the twentieth century the role of aesthetics in urban development in Chicago had moved from the macro-aesthetics of "big plans" to the micro-aesthetics of wrought-iron fences and planters. Mayor Richard M. Daley had visited Europe in the mid-1990s and appreciated details of landscaping like decorative fencing, urban trees, and flower boxes. Daley decided that this was what Chicago needed to save it from becoming just another declining Rust Belt city. He started by getting the city government to beautify city properties, streets, and parks with faux wrought-iron fences and plantings and then, in 1999, pushed through a City Landscaping Ordinance that required private businesses to spend their own money on such measures. Daley clearly shared Burnham's belief in the importance of beauty to urban life, but he sought to imprint his aesthetic vision on the city through a multitude of small plans.

The move toward small, incremental, additions to the city reflects the move from the grand modernist ambitions of urban renewal in the 1960s to the more piecemeal approach of TIF funding in the 1990s. The two are connected through the insertion into TIF agreements such as the one that includes Maxwell Street of a few lines specifying "planters, flower boxes, plazas, variety of tree species and wrought-iron fences where appropriate."

> The soaring ceilings, oversize windows and open space planning of traditional loft homes with all the high-tech, energy-efficient advantages of new construction. Every home features 11-foot ceilings, oak floors, a fireplace and an expansive, private balcony along with up-to-date necessities like pre-wiring for high-speed phone, data, and internet access.
> **CHICAGO'S NEXT GREAT NEIGHBORHOOD**[212]

Regulation/ Order

There is a long history of attempts to straighten out the market that combine aesthetics with regulation. The aesthetics of Maxwell Street were a constant source of contestation. To some, Maxwell Street's apparent disorder was appealing evidence of authentic urbanity. To others it was a thorn in the side, an impediment to the production of an ideal Chicago—a safe place for capital and the people who controlled it.

As early as the 1890s Chicago had attempted to erase the name Maxwell Street and replace it with West 13th Place. A strictly ordered street numbering system was preferred to the random names of an earlier time. Maxwell Street was to become part of the grid. Nobody paid attention.

In 1939 the Maxwell Street Merchants Association planned to modernize the market by insisting on a uniform size for street stalls and forbidding the shops along the street from using the sidewalk. Vendors' carts were to be painted orange and blue and covered with sanitary canvas awnings, with a garbage receptacle attached. The plans were greeted in the press by a litany of references to the olfactory chaos of the market.

> Its architectural ears will be scoured and its appearance and olfactory tempo vastly improved.
> ***CHICAGO DAILY NEWS*, 1939**[213]

> Smells? The modernizers look to a later day to begin the refining process on the Maxwellian potpourri. Later, too, they will essay revision of the street's cacophonous symphony of screeching wheels, barking dogs, wheedling voices, blaring radios and cackling, crowing and honking fowl.
> ***CHICAGO TRIBUNE*, 1939**[214]

Sound and smell once again signify the excessive. Like the market, they overspill boundaries, transgressing the limits of the proper. They have to be controlled or removed through the logic of planning.[215]

The plans for visual and olfactory uniformity were accompanied by a "code of conduct" that included the article "Maxwell Street no longer will condemn interventions, modernization and beautification without a careful examination."[216] If the idea caught on, it was only for a short time. There are no images of Maxwell Street with uniform carts in the Chicago History Museum archives.

In 1966 the city's Department of Urban Renewal included ideas for a new Maxwell Street Market in its proposals for the Roosevelt-Halsted area. The report reorganized the market as a special kind of place.

> The Maxwell Street Market is composed of business retail facilities in permanent structures and of merchants occupying sidewalk structures or temporary facilities erected for the weekend trade. This varied method of merchandising, plus the varied type of new and older merchandise offered has resulted in a unique market atmosphere. Such a market has developed in several of the central cities of the world's major metropolitan areas and has proved not only an excellent retail outlet, but a boon to the tourist trade as well. An opportunity for development of such a market in a new physical setting would prove an asset to the merchant, and to the customers seeking not only unusual merchandise, but the convivial atmosphere of the true open-air market.
> **CHICAGO DEPARTMENT OF URBAN RENEWAL**[217]

The "convivial atmosphere of the true open-air market" was clearly appealing to the writers of this report. They wanted it and yet did not want it. The sound, smell, and appearance of the market, the very things that had made the street such a rich place to other writers and photographers over decades, needed taming. Everything in its proper place.

> It is anticipated that the existing Maxwell Street open air market can be accommodated on the privately owned open areas related to this shopping center. The combination of permanent structures, plus temporary market facilities . . . would result in the provision of a colorful, festive atmosphere conducive to creating and retaining shopping potential.
> **CHICAGO DEPARTMENT OF URBAN RENEWAL**[218]

The proposals again stressed a need for uniformity, for everything "from sign lettering, to plaza pavement, to brick color and texture . . . to be utilized in harmonious and restrained manner." Again and again, planners and report-writers applaud the atmosphere of the market, then argue for ways to make this atmosphere more palatable, more "harmonious and restrained"—terms not often used to refer to the existing market. An ordered aesthetic based on standardized landscape features was consistently recommended.

In a 1966 visualization of a reformed Maxwell Street Shopping Center, stalls appear in neat, even rows within a contained courtyard, which is in turn surrounded by neat lines of evenly spaced trees. Trees, generally absent from images of Maxwell Street up to the 1980s, appear frequently in visions of the area's future.

The assemblage of sights, sounds, and smells that characterized Maxwell Street Market throughout most of the twentieth century was, in some ways, organic. The market became part of the city in a more or less spontaneous

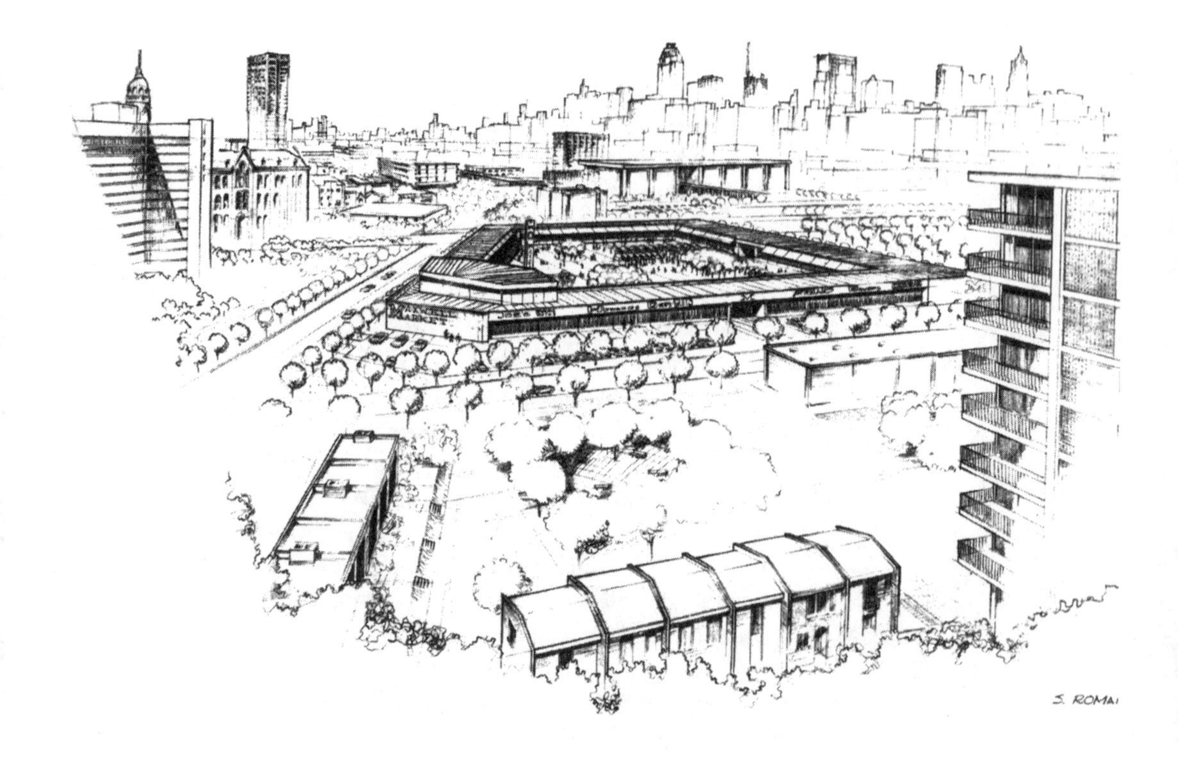

FIGURE 42.
Illustrative sketch of proposed shopping and residential area, Roosevelt-Halsted Proposals for Renewal, 1966.

way and maintained a more or less spontaneous form of order. The various schemes to order its excess sought to locate the market in a wider world of legibility imposed from outside. The abolition of smell, creation of uniform carts, and translation of the market into a "shopping center" were all part of this process.

> How did the state gradually get a handle on its subjects and their environment? Suddenly, processes as disparate as the creation of permanent last names, the standardization of weights and measures, the establishment of cadastral surveys and population registers, the invention of freehold tenure, the standardization of language and legal discourse, the design of cities, and the organization of transportation seemed comprehensible as attempts at legibility and simplification. In each case, officials took exceptionally complex, illegible, and local social practices, such as land tenure customs or naming customs, and created a standard grid whereby it could be centrally recorded and monitored.
>
> **JAMES C. SCOTT**[219]

Ordering Maxwell Street aesthetically was accompanied by a number of other forms of ordering. Up to the present day there have continually been questions about appropriate fees for stall-holders, payment of taxes, and standardization of measures and prices. Through most of Maxwell Street's history, all of these were negotiable. The market has its own *metis*, or practical knowledge. This was knowledge you learned in place through practice—the best ways to "cheat you fair."

> Any experienced practitioner of a skill or craft will develop a large repertoire of moves, visual judgments, a sense of touch, or a discriminating gestalt for assessing the work as well as a range of accurate intuitions born of experience that defy being communicated apart from practice.
>
> **JAMES C. SCOTT**[220]

> The "puller" is a specialist. He has developed a fine technique of blocking the way of passers-by. Before he is aware of it, the unwitting and unsuspecting customer is trying on a suit that is many sizes too large and of a vintage of a decade ago. The seller swears by all that is holy that it fits like a glove, that it is the latest model put out by Hart Schaffner & Marx, and that he needs money so badly that he is willing to sell it at a loss of ten dollars. If the customer is skeptical and is inclined to ask how the dealer can stay in business and lose ten dollars on a suit, he is told confidentially, "You see, we sell so many of 'em."
>
> On the sidewalk a puller shouts, "Caps, fifty cents!" In a moment he has a victim by the arm, and the salesman is trying on caps. "Yes, they are fifty cents apiece." He finds one that fits. "Seventy-five cents for that one."
>
> "But I thought you said they were fifty cents?"
>
> "Yes, but this one fits you!"
>
> **LOUIS WIRTH**[221]

The knowledge from the outside—the ordered and sweet-smelling aesthetics—corresponds to *techne*—systematic, abstract, often quantified and generalizable forms of knowledge. This is the knowledge of carefully measured carts, standardized fees, and externally imposed codified rules. *Metis* is local, *techne* is (or, more accurately, tries to be) universal.

The process of designating Maxwell Street as a TIF district was partly an aesthetic process that continued a long history of attempts at ordering and beautification.

717 West Maxwell Street

Should you take part in a conference at the University of Illinois at Chicago, you might be presented with a list of eating options in the surrounding area. You might decide to eat lunch at the WOW Café and Wingery, part of a chain based in Louisiana. As its name suggests, it specializes in chicken wings, which it serves with a variety of sauces, including the coconut curry "Bombay" and the chipotle chili "Santa Fe." WOW is located at 717 West Maxwell Street. The WOW Café and Wingery is part of the landscape of this place in Chicago. But what you see is not entirely what you get. The façade of this building used to belong to 1245 South Halsted Street. The old 717 West Maxwell Street façade now belongs to the ice cream shop next door at 719 West Maxwell Street.

FIGURE 43.
The 700 block of Maxwell Street, from the corner with Halsted, 2012. Photo by the author.

The building whose address was 717 West Maxwell Street was built in 1883 as a two-story structure. A third story and rear extension were added in 1909. The extension was designed by the architect David Saul Klafter. Klafter began his career as an office boy for the better-known Chicago architect Louis H. Sullivan, and later worked for the prestigious firm D. H. Burnham & Company (1907–1908). After starting his own practice in 1911, Klafter was responsible for many substantial commercial and residential buildings throughout the Chicago area. These include some of Chicago's first cinemas as well as the former Trailways bus terminal in the South Loop, where you can find Klafter's name inscribed in the terra-cotta façade.

We do not know much about what happened in 717 West Maxwell Street in the years after 1909. We know that originally it did not have a storefront but that one was added. We know that the ground floor was home to a number of businesses—including Maxwell Street Meat Market, Frank's Lamp Shop, and a fish market—while the upper floors were rented out as residential units.

The practices that went on in 717 West Maxwell Street and meanings associated with it over nearly a hundred years are more or less invisible to us. Its materiality, however, continues to matter. Klafter's addition to the building included several features that made it architecturally interesting.

> On the west side of the façade there is a set of pilasters that extend from the sign frieze to the lintel level of the third story. The pilasters frame a triple window at the third story. There is also a single rectangular window on the east side of the facade. That window opening is now filled in with brick. Both the single rectangular window opening and the triple window opening on the

second story are also now filled in with brick. Above the third story window openings, at what would be the lintel line, there is a limestone belt course with molded coping. Above this is a frieze of patterned brickwork. Above this is a pediment with limestone coping over the bay framed by the pilasters. At the flat east side of the façade, the limestone coping of the pediment extends to a belt course separating the frieze above the third story from the parapet above it.

NATIONAL REGISTER OF HISTORIC PLACES REGISTRATION FORM, 2000[222]

An archaeological survey of the Maxwell Street area conducted in 2000 concluded that the façade of the building was worth saving from demolition. That is why the façade was moved to its current position.

While carefully leafing through images of Maxwell Street in the Chicago History Museum with my white cotton gloves I kept an eye open for an image of 717 West Maxwell Street, hoping to get a sense of the energy of life that surrounded that site in the market's heyday.

The building has a ghostly presence in the distributed archives. In the UIC University Archives, in files labeled "Associate Chancellor, South Campus Development Records, 003/02/02, series 1, box 28, file 28-225," I found a record of the purchase of 717 West Maxwell purchase from June 29, 1992. UIC bought the property, described as a three-story brick building with no written leases (there were occupants on month-to-month tenancies). from William Shapiro. On the same day there is a letter from George W. Davis of the law firm Schiff Hardin and Waite to Kenneth M. Smyth, the associate university counsel, stating Shapiro's properties had been appraised and that 717 West Maxwell had been valued at forty thousand dollars. On August 27, 1992, Davis wrote to Ann Marie Lodl, executive assistant of the Office of the Vice Chancellor for Administration, enclosing a copy of the appraisal conducted by Urban Real Estate Research, Inc.

The subject property consists of a rectangular shaped parcel, zoned B3-4, General Retail District containing approximately 2,250 square feet improved with a three story masonry constructed commercial building with storage rooms on the two upper floors. The building contains a total of approximately 5,750 square feet of area. The parcel is located on the south side of Maxwell Street south of Roosevelt Road approximately one half block east off Halsted Street.

The structure is only fair condition and would need a large amount of money invested into it to bring it back up to the caliber of the properties along Halsted Street and even more to make the building compatible in condition to those buildings located in similar university neighborhoods to the north of the subject area near Taylor Street and Racine.

APPRAISAL OF 717 WEST MAXWELL STREET, 1992[223]

The lot is formally located and identified

FIGURE 44.
Maxwell Street, crowd on sidewalk, 1930. Photo by Chicago Daily News. *DN-0091669. Courtesy of Chicago History Museum.*

> Lot 7 of J. Nutt's Subdivision of Lots 1, 2 and 3 of Block 64 of Canal Trustee's New Subdivision of Blocks in the Northwest 1/4 of Section 21, Township 39 North, Range 14 East of the Third Principal Meridian, in Cook County, Illinois.[224]

The surrounding neighborhood is described.

> This smaller, immediate area is in much worse condition than the rest of the surrounding area. The vast majority of the structures in this area are old and very much deteriorated. Much of the land on the subject's block is vacant and strewn with garbage and makeshift housing for homeless people. Typical structures on the subject's block include single and multi-story commercial and residential buildings with gates or roll down doors for protection and abandoned structures in very poor condition.
>
> The streets are also littered with garbage and abandoned cars.[225]

The condition of the building is assessed.

> The building was built in 1889 and received little maintenance over its life and is in overall poor to fair condition.

Exterior Walls: The outside of the building is a brick finish, with some decorative stone features at the front around the windows and doors. . . .

Floors: The floors are of a common wood and have not been kept up showing signs of warping and water damage.

As mentioned, this building is in only poor to fair condition and appears to have been this way for a number of years. The interior of the building is for the most part greatly deteriorated. The plaster on the walls has fallen away in many areas exposing the interior sections of the walls, the electrical wiring and the lath. A significant amount of money would need to be invested to repair the walls. The upper floors of the building can only be accessed from stairways starting on the first floor of the building along the west wall. The second and third floors have no doorways or formal rooms which could be rented out to residents but rather are simply moderately sized rooms used to store items which would normally be found at the weekend flea markets. The exterior of the building is also in only fair condition. Many cracks have developed in the brickwork along all elevations. The upper floors of the northern elevation which is the storefront of the building have been boarded up and locks have been installed to keep out vandals and thieves. The use of different types of material and the deterioration of the buildings and these materials has yielded an unappealing structure.[226]

The report is accompanied by photographs of the front and back and third-floor interior showing piles of junk—a cage overflowing with papers.

This is the surviving archival trace of 717 West Maxwell Street. A few scattered fragments.

A few more are recorded in the "Maxwell Street Archives and Artefact List" compiled by members of the Maxwell Street Historic Preservation Coalition.

French-fry maker (manufacturer: Bloomfield Industries, Chicago) (Collected by MM)
Document: "Order in the Circuit Court of Cook County, Illinois," no. EOL50734 dated 10/1/2002, for Zafar Sheikh
One fragment of awning canvas, green, approx. 9" × 7½" (Collected by LG, 02)
MAXWELL STREET ARCHIVES AND ARTEFACT LIST

These are things collected from the building that was 717 West Maxwell Street before its façade was removed and the rest demolished. They were collected by preservationists as the landscape was erased.

Trees/Race

On one of my walks about the campus of UIC I noticed a sign informing me that a forest was growing.

An inventory was completed of UIC's approximately 5,000 trees in the summer of 2011 using i-Tree, a program developed by the United States Forest Service.

i-Tree calculates the benefits of trees, including carbon sequestration, pollutant removal and monetary value. The inventory is updated yearly. The entire campus tree inventory and benefit data is available for UIC students and faculty to use.[227]

The UIC campus forest includes more than a hundred species and is linked to the university's "Climate Action Plan." It is also part of the "Tree Campus Network," sponsored by the Arbor Day Foundation and Toyota. Maps of the forest show hundreds of trees in the area where the market used to be.

The forest is part of an area (the Near West Side) where, between 1980 and 2000, the African American population decreased from 72.2 percent to 52.9 percent. In the census tracts for what is now called University Village, the African American population was 15 percent in 2013. There is a longstanding association between "green space" and the "improvement" of the poor. Green space, it has long been argued, both increases property values and encourages moral behavior. It has also been argued that urban forest canopy cover is unevenly distributed in relation to race. White people get more trees.[228] In the case of the UIC campus, the arrival of trees coincides temporally with decreasing numbers of black people and increasing property values.

On and around Maxwell Street there are now plenty of trees, plant boxes, and short faux-wrought-iron fences. There are trees everywhere in University Village and on the UIC campus. They even appear on maps of the college plans. This material (and imagined) landscape is clearly implicated in a process of urban transformation that produces new kinds of value.

There are two clear narratives in this process that work through the material landscape. The first is the narrative of "blight," which draws on botanical pathology to condemn landscapes viewed as derelict and decayed and to suggest that this decay poses a threat of spreading. The second is the narrative of beautification, symbolized and then materialized by trees, decorative fences, and plant boxes. These narratives are intertwined with both old and new landscapes as well as with the political-economic instrument of TIF funding in the city.

Blight

To be considered "blighted" an area had to have little chance of being "developed" without action by the city and to exhibit five or more attributes from a list of factors deemed "detrimental to the public safety, health, morals, or welfare."

Age; dilapidation; obsolescence; deterioration; illegal use of individual structures; presence of structures below minimum code standards; excessive vacancies; overcrowding of structures and community facilities; lack of ventilation, light or sanitary facilities; inadequate utilities; excessive land coverage; deleterious land use or layout; depreciation of physical maintenance; or lack of community planning.[229]

FIGURE 45.
Map of "age" factors in Roosevelt-Union TIF district.

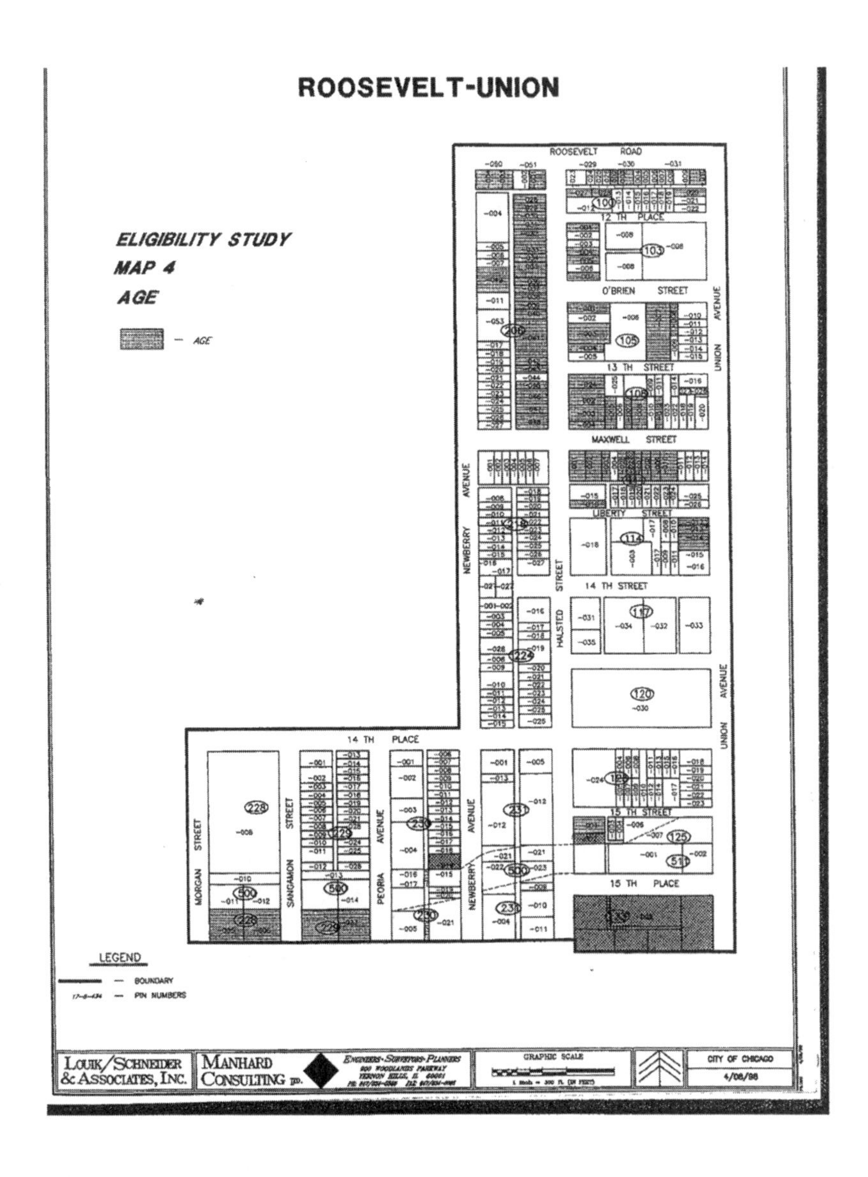

Nine of these were discovered in the Roosevelt-Union area, including, to a major extent, age, dilapidation, deterioration, excessive vacancies, excessive land coverage, and depreciation of physical maintenance. Each factor was then separately mapped. The definition of "age" as a factor in blight reads:

> Age presumes the existence of problems or limiting conditions resulting from normal and continuous use of structures which are at least thirty-five (35) years old. In the Redevelopment Project Area, age is **present to a major extent** in sixty-seven (67) of the seventy-three (73) (ninety-one and seven-tenths percent (91.7%)) buildings and in thirteen (13) of the sixteen (16) (eighty-one and three-tenths percent (81.3%)) blocks in the Redevelopment Project Area.[230]

The Roosevelt-Union Redevelopment Plan and Project concluded that the redevelopment of the area would cost $103 million. The baseline equalized assessed value (EAV) was set at $3,968,563, with an expected EAV for 2008 of $48 to $55 million.

Once again, the act of mapping and defining forms part of the quasi-scientific process of stigmatizing an area, which is a necessary precursor to the revaluing process.

> Municipalities justify such interventions by strategically stigmatizing those properties that are targeted for demolition and redevelopment. These justifications draw strength from the dual authorities of law and science in order to stabilize inherently ambiguous concepts like blight and obsolescence and create the appearance of certitude out of the cacophony of claims about value.
> **RACHEL WEBER**[231]

The negative attribution of "age" to property contributes to a definition of obsolescence—the notion that something is no longer of the times—is out of date. "Age," in this sense, is of course not susceptible to scientific diagnosis. In another context, that of historic preservation, age is a positive benefit. In the context of diagnosing blight, however, a designation of age ties a narrative of lack of value into the material landscape in order to legitimate practices of demolition and redevelopment that suit the purposes of private capital. Narrative again becomes part of the landscape.

The appraisal of the Roosevelt-Union TIF district includes a map for each criterion for blight, as well as matrices codifying the presence of the various criteria block by block (fig. 46). In this way, the City of Chicago draws on the apparently objective and scientific power of cartography and grid matrices in order to make inherently unstable concepts such as "age" appear to be verifiable fact. These practices seek to impose certainty on a landscape where notions of value are contested and uncertain.

The process of TIF designation around Maxwell Street met with protest and resistance. Steve Balkin, an economist from Roosevelt University, located in the Loop, penned a series of press releases on behalf of the Maxwell Street Preservation Coalition pointing out some of the gaps in TIF reasoning. The "age" designation was one of his targets for ridicule.

> This area is historic and research shows that people like to live in and near historic districts and are willing to pay more for property for that. In this case, "Age" of buildings is not evidence of "blight to a major extent." Instead, it is evidence that this area is historic and that its historic buildings should be preserved as part of any plan that purports to be promoting development.
> **STEVE BALKIN**[232]

> A Tax Increment Financing (TIF) district is suppose [*sic*] to lead to creation of jobs and revitalization for blighted communities, communities unlikely to improve in the future through free market forces. But most TIFs, as used in Chi-

FIGURE 46.
"Distribution of Criteria" matrix for part of Roosevelt-Union TIF district.

974 JOURNAL--CITY COUNCIL--CHICAGO 5/12/99

(Sub)Exhibit 3.
(To Roosevelt/Union Tax Increment Finance Program Eligibility Study)

Distribution Of Criteria Metrix.

B.

Improved Portion Of The Study Area.

BLOCK	1	2	3	4	5	6	7	8	9	10	11	12	13	14
17 20 206	X		X	X			X				X	X	X	
17 20 218														
17 20 224														
17 20 228	X	X	X	X			X				X	X	X	
17 20 229	X	X	X	X			X				X	X	X	
17 20 230	X	X	X	X			X				X	X	X	
17 21 100	X	X	X	X							X	X	X	
17 21 103	X	X	X	X			X				X	X	X	
17 21 105	X	X	X	X			X				X	X	X	
17 21 108	X	X	P	X			X				P	X	X	
17 21 111	X	X	X	X			X				X	X	X	
17 21 114	X		P	X									X	
17 21 117	X		X	X									X	
17 21 120				X									X	
17 21 125	X	X	X	X									X	
17 21 133	X	X	X	X			X						X	

Key
X Present to a Major Extent
P Present
Not Present

Criteria
1 AGE
2 DILAPIDATION
3 OBSOLESCENCE
4 DETERIORATION
5 ILLEGAL USE OF INDIVIDUAL STRUCTURES
6 PRESENCE OF STRUCTURES BELOW MINIMUM CODE
7 EXCESSIVE VACANCIES
8 OVERCROWDING
9 LACK OF VENTILATION, LIGHT OR SANITARY FACILITIES
10 INADEQUATE UTILITIES
11 EXCESSIVE LAND COVERAGE
12 DELETERIOUS LAND USE OR LAYOUT
13 DEPRECIATION OF PHYSICAL MAINTENANCE
14 LACK OF COMMUNITY PLANNING

cago, including the Roosevelt-Union TIF for the old Maxwell Street area, are, instead, a tool used by real estate developers to displace existing residents and businesses. TIFs in Chicago are not about development. They are about displacement.

In the old Maxwell Street area, UIC and its real estate development partner, Mesirow Stein Real Estate Inc., will be using TIF funds to destroy historic buildings, demolish the remaining hot-dog stands (Jim's Original is in a building built before the Chicago Fire), and kick out the merchants and residents, many of whom will likely become homeless. TIF funds will enable the construction of

$300,000 townhouse condos right next to Pilsen, by the viaducts. The placement of those townhouses are expected to dramatically accelerate gentrification in Pilsen.

STEVE BALKIN[233]

Balkin and others argued that TIF funding was a process that deprived local school and park districts, for instance, of much needed funds in order to provide a subsidy to private real estate developers. Despite the talk of "revitalization" of deprived neighborhoods, they argued, the process most often leads to profits for absentee landlords and nice homes for relatively well-off incomers.

Maps/Territorialization/Deterritorialization

The maps that placed the intersection of Maxwell and Halsted first in an urban renewal zone and then in a TIF district perform territorialization. They draw a thick black line around a space and define a number of practices that can or cannot go on within it. They also draw on a lexicon of meaning to define the area at a particular point in time and attempt to create a new set of meanings for the future place.

But these maps, and the definitions that accompany them, are simultaneously deterritorializing tools that bring the landscape into conversation with other places elsewhere. The urban renewal map of 1966 connects the area to a lexicon of blight and improvement, and to a flow of federal tax funds. With the TIF map, the deterritorializing effects are even more remarkable: the local state creates a territory that financializes the landscape, making new forms of "capital switching" possible. "TIF is the governance mechanism . . . that allows the income stream (property tax revenues) generated from locally embedded assets (property) to be converted into financial instruments and exchanged in the global marketplace."[234]

In other words, the map is part of the process by which an area is locally defined as failed and obsolete in order to enact the securitization of real estate. For bits of the local landscape to enter expansive and fluid global bond markets, which are always elsewhere, the local place must be dematerialized and deterritorialized, stripped of specificity and made increasingly abstract.

The Maxwell Street Market was defined as part of such a "blighted" landscape, and the TIF process was just the latest in a series of renewal processes that threatened its survival. Writers of a 1980 report on the future of the market were keen to see it survive.

> The historical market must be retained.
>
> Maxwell Street is a unique and resilient institution faced with recent abuse which can survive on its own momentum as it has existed for 120 years—the span of several lifetimes. The area shares, even while whole buildings are physically depleted and lots are in terrible condition, a strong mercantile history that has a spiritual and cultural presence.
>
> The ethnic pluralism of the Maxwell Street Market is precious, and the sense

that this neighborhood is not the turf of any one ethnic or economic group in particular makes it similar to State Street as a universal neighborhood where people from many backgrounds can feel at home. This cosmopolitan aspect should not be lost by the City. It is educational, entertaining, and teaches the tough lessons of give and take that form much of what one practices in commerce above the retail, fixed and computerized-priced suburban experience.
CHICAGO DEPARTMENT OF PLANNING, 1989[235]

In 1994 the market was formally moved to Canal Street. In 2008 it was moved again, to its present location on Desplaines Street.

Historic Preservation

The Maxwell Street Historic Preservation Coalition attempted to create yet another territorial definition for the area around Maxwell and Halsted. And like the City of Chicago, it produced a map and boundary definition.

These acts of representation attempted to enact the formation of a preservation district—an entry on the National Parks Service's Register of Historic Places. Had they been successfully woven into the landscape, they would have had material effects. The City of Chicago and UIC would have been unable to demolish the buildings contained within the specified boundary. In this sense, the coalition's proposed preservation district was the inverse of the urban renewal areas successfully defined by the City of Chicago in the 1960s.

The Maxwell Street Historic Preservation Coalition formed to prevent the erasure of the market and the transformation of the area around Maxwell and Halsted to the upscale housing development known as University Village (and later, University Village Maxwell Street). While its efforts largely failed, a few storefronts were preserved. Archaeologists who assessed the area for potential listing on the National Register of Historic Places deemed the façade of 717 West Maxwell, along with a number of others, to be a "contributing object"—a structure with interesting architectural features that remained largely "original" and "unchanged." It had "integrity" and was therefore granted the right to exist. The objects gleaned from the property by coalition members could not be placed in or around the new façade. They were taken to people's homes. They were valued. One day, if all else failed, it was hoped that they would become part of a museum commemorating the market.

As UIC moved in and threatened to demolish the area around the market a group of local market enthusiasts, including the coalition, blues musicians, residents, and stall-holders instigated a highly visible grassroots campaign to save the market and surrounding area. The campaign included street theater, blues concerts, persistent letter writing, and awareness-raising. Pamphlets, flyers, and notes handed out on the street emphasized the history of the site and encouraged the university to reverse its planned leveling of the area.[236]

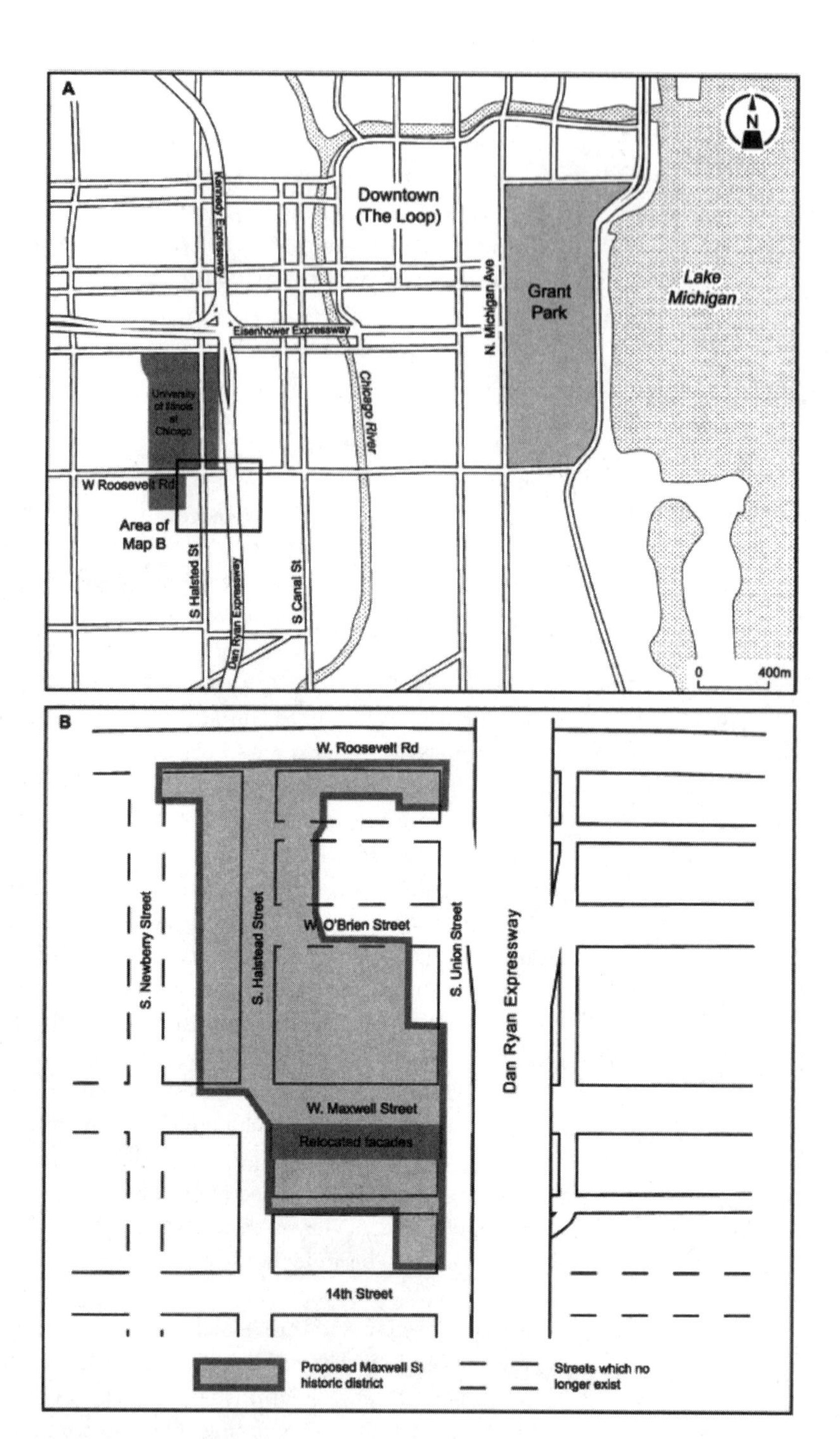

FIGURE 47.
Map of proposed Maxwell Street Historic District and corresponding textual definition. Map prepared by Jenny Kynastone.

The boundaries of the Maxwell Street Historic District begin at 701 W. Roosevelt Road at the southwest corner of the intersection of Roosevelt Road and Union Street; west on the south curb line to the northwest corner of the Church of St. Francis of Assisi property at 817 W. Roosevelt Road; south on the east curb line of Newberry Street to the alley that runs along the building's south lot line; east on the south curb line of the alley to the north-south alley located mid-block that borders the rear of all buildings on the west side of Halsted Street, starting at 1212 S. Halsted; south on the west curb of the alley to Maxwell Street; east on the north curb line of Maxwell Street to the southwest corner of 801 W. Maxwell; southeast (diagonally) to the southeast corner of 743 W. Maxwell and continuing south on the east curb line of Halsted Street to the south curb line of Liberty Street; east on the south curb line of Liberty Street to the west lot line of lots facing Union Street; south to the southwest corner of 1352 S. Union; east to the southeast corner of 1352 S. Union; north on the west curb line of Union Street to the north lot line of 1300-1304 S. Union (at W. 13th Street); east on the south lot line of W. 13th Street to the east curb line of the alley (bordering 711 W. O'Brien); north on the east curb line of the alley to the south curb line of W. O'Brien Street; east on the south curb line of W. O'Brien to 741 W. O'Brien and north (diagonally) to the east end of 1237 S. Halsted; north following the rear property line of buildings on the east side of S. Halsted Street to the south curb of the alley immediately south of Roosevelt Road; east on the south curb line of the alley to the west lot line of 1210 S. Union; south on the west lot line to the south side of the lot line and east on the south lot line to the west curb of Union Street; north on the west curb of Union Street to the point of beginning at 701 W. Roosevelt.

On the back of a T-shirt commemorating the centennial of the market was a list of Maxwell Street blues singers:

Papa Charlie Jackson
Muddy Waters
Jim Brewer
Little Walter
Robert Nighthawk
Bo Diddley
Arvella Gray
Jimmie Lee Robinson
Carey Bell
Jimmy Rogers
Moody Jones
Johnnie Mae Dunson
Michael Bloomfield
Daddy Stovepipe
Jimmy Davis
John Wrencher
Johnnie Young
Playboy Venson
Robert Whitehead

Since you've started with this damn wrecking ball
our lives, our dreams, our needs continue to fall.
The great great history of this area
which one stood proud and tall
guess what UIC
you've managed to destroy it all.

JIMMIE LEE ROBINSON, "MAXWELL STREET, TEAR DOWN BLUES"

> Maxwell Street was the perfect setting for the Blues. It was the dirtiest, most depressed part of the city. The boarded-up buildings and broken sidewalks were the real backdrops. You can see how songs like Little Walter's "Blues with a Feeling" and "Mean Old World" were born out of this atmosphere.
> ***CHICAGO BLUES GUIDE***[237]

In 2000 the blues singer Jimmie Lee Robinson, disgusted with the demolition of buildings in the Maxwell Street area, went on hunger strike for eighty-one days.

Materiality/ Memory

Activists formed the Maxwell Street Historic Preservation Coalition to work toward placing the area on the National Register of Historic Places—an act that would have made if considerably harder for UIC to demolish its structures. The application process required a formal archaeological survey to establish which buildings were worthy of preservation—worthy of being designated, in National Parks Service jargon, a "contributing object." This formal survey can be read alongside a series of press releases and calls for action from the coalition and other activists who were seeking to save the material structure and character of the area around the intersection of West Maxwell Street and South Halsted Street in Chicago.

The copresence of stability and change in a place like Maxwell Street makes place's connection to memory and the past all the more complicated. The materiality of place has the most obvious connections to memory and memorialization. The material nature of buildings and roads and passageways means that they endure—not forever, perhaps, but for considerable passages of time. This endurance provides an anchor for stories that circulate in and around a place. It reminds us of things.

> As a geographer, I could not help but notice that the sites themselves seemed to play an active role in their own interpretation. What I mean is that the evidence of violence left behind often pressures people, almost involuntarily, to begin debate over meaning.
> **KENNETH FOOTE**[238]

The experiential nature of place adds to the association between place and memory. It is one thing to have a place represented to us in a novel or a painting or a museum; it is quite another to actually be in the place. While the materiality of place may endure, the experience of place is never the same twice.

The procedures through which properties are entered into the National Register entail a surfeit of nomination forms, evaluations, assessments, and correspondence, each affording a glimpse into the way the vagaries of place are negotiated. Central to this negotiation process are highly normative determinations regarding space and time—the duality of place, its obduracy and its changeability—which serve to frame properties within a consistent historical period and contiguous geographical area. In the case of Maxwell Street, the place proved to be too promiscuous—too obviously an assemblage—to be so constrained.

As it turned out, a few pieces of area architecture were "preserved." In an attempt to pacify those objecting to the destruction of the market area, Mayor Richard M. Daley instructed the university to preserve thirteen façades. These were relocated to a single block to act as a symbol of the area's past. The façade of 717 West Maxwell Street, as we've noted, moved next door, still on the south side of Maxwell Street. The façade to its left belonged to 727 West Maxwell Street, and the one on the right is from 1245 South Halsted Street. Across the street is the façade of 722 West Roosevelt Street. Behind these façades are restaurants, coffee shops, and a multistory parking lot.

The Maxwell Street Historic Preservation Coalition sought to have the Maxwell Street area placed on the National Register of Historic Places twice, in 1994 and 2001. The first application highlighted its history as a marketplace and its association with, particularly, Jewish immigration. The second emphasized the area's history as an entertainment space—as the birthplace of the blues. Both linked processes of migration to more local practices (market selling and music). And both failed. These failures were based on (mis)understandings of what gives a place integrity. For a place to be deemed worthy of preservation by the National Park Service, it must have "integrity," a notion based largely on the materiality of place and its capacity to resist change.

Applicants for historic-place status in Illinois have first to submit a portfolio of information to the Illinois Historic Preservation Agency (IHPA), which then returns advice regarding the likely success or failure of the bid and how it might be improved. In the case of Maxwell Street, the IHPA twice assessed the nomination favorably, providing the necessary application materials along with advice on the procedures involved.

The Maxwell Street Historic Preservation Coalition then completed the National Register of Historic Places application and returned it to the IHPA, which brought the application to a meeting of the Illinois Historic Sites Advisory Council, where the applicants and those supporting and contesting the nomination were given an opportunity to make their views known. At the end of the 1994 meeting, the council voted unanimously to approve the nomination to the National Register. The IHPA then prepared a fifty-two-page preliminary opinion, accompanied by photographs and details of specific properties in the area. A copy of this report was sent to the Landmarks Division of Chicago's Department of Planning and Development, and to Mayor Daley, for comment.

Key to the IHPA decision was that the area "retains sufficient integrity of location, setting, feeling, association, and materials from the period of significance."

> The buildings, often nondescript, which contained the storefronts and the apartments of their proprietors, and the streetscapes crowded with pushcarts and stands were the physical context in which the bustling commerce and acculturation took place. The historical significance of the area lies not in the occurrence of particular events of note within its confines, but in the vitality that took place from day to day in the area.
> **ILLINOIS HISTORIC PRESERVATION AGENCY, 1994**[239]

The principle case for historic-place status, then, reflected its intangible, slippery, mundane, and everyday qualities. Maxwell Street was a fluid place that was experienced in a myriad of ways. It was the conflict between its material "persistence" and its more transitory qualities that became the center of debate about the status of Maxwell Street.

The Chicago Landmarks Commission (CLC), following a meeting on May 4, 1994, concluded that, except for an old police station that was included in the wider area, Maxwell Street should not be nominated to the National Register.

In a letter to Ann Swallow of the IHPA, Peter Bynoe of the Chicago Landmarks Commission explained the CLC's view that although the area certainly had "rich historical associations," it did not have "sufficient integrity" to convey this history, as much of the material structure no longer existed. He argued that the boundaries of the proposed historic place had no apparent purpose "other than to include every remaining structure in the area," many of which had, in any case, been altered in ways that undermined their ability to convey the history of the area. Most tellingly, perhaps, the letter made a distinction between the activities that had taken place in the area and the material structure of the place itself.

> The street activity of the Maxwell Street Market is its primary historical association. Given the makeshift, transitory nature of this activity, the buildings alone cannot convey the historical feeling of the market. The National Register assists in the preservation of buildings, not such fluid activity as was the historical essence of Maxwell Street.
> **COMMISSION ON CHICAGO LANDMARKS, 1994**[240]

As far as the CLC was concerned, the Maxwell Street preservationists had failed to tie the material structure of the market area to the historically significant but "fluid" activities that took place there. This link between material landscape and practice was crucial.

Finally, the application was forwarded to the keeper of the National Register of Historic Places in Washington, DC, along with the positive IHPA report, the contrary CLC assessment—and a letter from William Wheeler, the Illinois historic preservation officer, who, concurring with the CLC's judgment, offered his own negative assessment.

Integrity/ Persistence/ Practice

The issue of the fluid nature of the events that took place at Maxwell Street arose repeatedly throughout the process of applying for inclusion in the National Register for Historic Places. It was obviously on the minds of some of the people who wrote letters in support of the application.

> Regarding the integrity issue, one must look beyond the loss of structures into the fabric of the function of the market. Buildings are not what makes Maxwell Street historic; street vendors, musicians, and ambience did and [do]. The complex bringing together of persons from all walks of life—poor, rich, black, white, young, old, Hispanic, Jewish, eccentric, intellectual—is what makes Maxwell Street historic. The continuous use of this same place . . . as a poor person's business incubator for 125 years is what makes Maxwell Street historic.
>
> **WILLIAM GARFIELD, 1994**[241]

> We believe that the proposed district still has the feel of a 1920s–30s commercial district. Despite the not-so-benign neglect of the city and university's effort to clear the area for its own use, you can walk around Maxwell and Halsted Streets and feel you are in a bygone era. . . . It is important to remember that the Maxwell Street Market during much of the historic period was a crowded, busy, noisy, messy and (in the earliest period) often muddy set of streets. It never resembled today's malls or metropolitan boulevards with boutique shops and upscale restaurants.
>
> **LORI GROVE AND ELLIOT ZASHIN TO IHPA, JUNE 9, 1994**[242]

The appeal to "feel[ing] you are in a bygone era" suggests that the material structure of the Maxwell Street area could impel the present visitor to imagine being corporeally there in the past. One way in which places compel engagement is through the interaction of material structure, representation, and practice. Not simply things to look at, places provide the context for existence. We have to move through them and embody them. Whereas a book can be read, a place has to be inhabited.[243]

Their link to practice makes places particularly powerful ideological tools as they link the things that are represented by material structure to the often unconscious realm of habit. This also makes them powerful repositories of memories that can be reactivated through embodied practice. This is why, Delores Hayden argues, New York's Lower East Side Tenement Museum works so well. You are not simply seeing pictures of abject poverty or watching a film or reading a pamphlet—you are in the place. You are compelled to engage.[244] The Maxwell Street advocates argued that such an engagement could still be compelled precisely because of the messiness of the area. The Chicago Landmarks Commission and others countered that a different kind of complete landscape was necessary to compel such an engagement.

Despite the unanimous verdict of the IHPA that the Maxwell Street district deserved historic-place status, the state's historic preservation officer (who was responsible for forwarding the nomination to Washington) concluded that it did not, due to the lack of integrity of setting, design, and materials. In his view, there simply was not enough left of the material landscape that had existed in the "period of significance" (1880–1944). His letter focused on the extensive renovations to extant buildings, as well as the demolition of large numbers of buildings within the district. Noting the high level of popular support for the nomination, evidenced by many supporting letters, he wrote that he did "not believe that the integrity issue has been considered or given sufficient weight by most people."[245]

Carol Shull of the National Park Service wrote to Lori Grove of the Maxwell Street Foundation in December 1994 informing her that "due to the irretrievable loss of historic integrity, the Maxwell Street Market Historic District does not meet the National Register Criteria for Evaluation and thus is not eligible for listing in the National Register."[246] The character of the district, she continued, "has been destroyed over time by demolition, neglect, and alteration. The district as a whole no longer retains the requisite qualities of design, materials, workmanship, and setting to qualify for listing in the National Register."

The Maxwell Street Historic Preservation Coalition tried again to get the area on the National Register in 2000, shifting the focus of the application from commercial history to the area's entertainment history as the birthplace of the Chicago blues. Once again, the application failed due to lack of integrity. On this occasion William Wheeler's letter to Carol Shull was even more damning, pointing out that the continued demolition of the area by UIC made it even less suitable for historic-place status. He commented not only on the lack of integrity of the material aspects of place but also on the more subjective "integrity of feeling."

> I find it difficult to believe that most persons visiting Maxwell Street today will find that Maxwell Street conveys feeling that is similar to that which would have been conveyed during the period of significance. The simple fact is that too many buildings are entirely missing or severely modified. . . . It is not enough for a district to contain scattered buildings that date from the period of historical significance and which still display some architectural features from the period. To qualify under the National Register's district definition of "a significant concentration, linkage, or continuity," the whole must transcend the sum of the individual pieces.
>
> **WILLIAM WHEELER, 2000**[247]

The fixation on material objects as indicators of "integrity" led to some discussion about the presence of wooden market stands on Maxwell Street, which the preservation coalition argued conveyed something of area's association with its historic commercial function. The narrative description of the area on the 1994 application form claims that some of these wooden stands should be included as contributing objects. In her initial response to the first draft of the application, Ann Swallow of the IHPA writes:

> You wrote that the information on the street stands is from correspondence—with the owner of the stand or neighborhood people? Are they saying that there has always been a stand at this particular location since 1890 or 1940? Or are they saying that the actual physical building standing today is what was built in 1890 or 1940? I think this would be difficult to prove. These stands are not built as permanent buildings—I'm sure they are completely rebuilt from top to bottom over the space of 15 to 20 years. If you cannot verify the date of the actual materials, then I would recommend not including them in the resource count.
>
> **ANN SWALLOW, 1994**[248]

The stands remained on the final registration form but were accompanied by a note from Swallow to the Chicago Landmarks Committee.

> In my opinion the wooden stands on Maxwell Street are not contributing historic resources. It is acknowledged that historically stands and tables were an important visual component of the market area, however, there is no evidence that the building fabric of these structures standing today date from before 1944.
>
> **ANN SWALLOW, 1994**[249]

> The ship wherein Theseus and the youth of Athens returned from Crete had thirty oars, and was preserved by the Athenians down even to the time of Demetrius Phalereus, for they took away the old planks as they decayed, putting in new and stronger timber in their places, in so much that this ship became a standing example among the philosophers, for the logical question of things that grow; one side holding that the ship remained the same, and the other contending that it was not the same.
>
> **PLUTARCH**[250]

The case of the wooden carts on Maxwell Street present an ideal case of what Manuel DeLanda refers to as assemblages—wholes whose "properties emerge from the interactions between parts."[251] In this sense the carts are representative of all places.

The principle case for historic-place status was the area's intangible quality of everydayness—the very kinds of things that were evident in over a century's worth of newspaper accounts. Maxwell Street was a fluid place that was experienced in myriad ways.

The members of the Maxwell Street Historic Preservation Coalition failed in their attempt to block the UIC-backed development of University Village, which has now taken over the area. The immediate cause of their failure was the transient nature of the material landscape. Yet the documents they compiled in nominating Maxwell Street for the National Register of Historic Places are themselves worth examining. These include a forty-four-page list of buildings in the proposed district, along with text drawn from an archaeological and historical evaluation of the area, conducted by independent researchers, which distinguished between "contributing" and "non-contributing" objects on a building-by-building basis. In each case the material structure of the building was assessed in terms of what it could contribute to the "integrity" of the place. The building at 717 West Maxwell was considered a contributing object. The one at 1310–1316 South Halsted received a different assessment.

> This one story brick building was constructed in 1883, and has had numerous alterations over the years. In 1912, architect Alexander Levy was hired by then owner William Franski to add one story to what was then a two story building, and convert the structure to include a theatre. In 1931, there was a permit to build an addition to the picture booth. In 1935, the two top stories were removed and alterations were made to the storefront. While some of the historic fabric likely remains, today the building is clad with vertical metal siding and the storefront is recent construction. There is no evidence that an intact historic façade exists beneath the siding. As the building has no historic integrity, it would likely be deemed as a non-contributing feature to a potential historic district.
>
> **NATIONAL REGISTER OF HISTORIC PLACES REGISTRATION FORM, 2000**[252]

In total the report identifies forty-four "contributing features" and ten "non-contributing features." Each assessment is based on the degree to which the material fabric of the place is as it was in the "period of significance." The changes made to 717 West Maxwell over the years were not deemed to have compromised its integrity, while the material transformations of 1310–1316 South Halsted rendered it "non-contributing." Somewhere between them is a point at which the degree of change switches from acceptable to not acceptable. Despite the fact that all places, and all material objects, are fluid and changeable, historical significance is attached in this reasoning to objects that are perceived to have stayed the same. Their "persistence" becomes their most important characteristic. Mobility in time and space is conceptualized as the enemy of integrity.

Activists and proponents of preservation in Maxwell Street emphasized the vitality of a living landscape in their efforts to get some kind of protection for the area, while the officials responsible for making decisions emphasized unchanging materiality. But while relative material constancy was key to the preservation of a number of buildings and façades in the Maxwell Street area, it would be wrong to suggest that materiality itself required preserving. Maxwell Street, as we have seen, was positively overflowing with things—some of them goods and some of them garbage. The question is, how do some things persist while others are discarded?

Consider a list of things gleaned from 1310–1316 South Halsted.

> Building inspection certificate for 1314½ (S. Halsted), dated 1964
> Colt 45 plastic and paper sign with 3-D bottle, 15″ × 10″
> "Only Remy," tin sign, 10″ × 17″
> 1 Marlboro race car illuminated electric sign, 31″ × 18″ × 2″
> Misc. liquor store signs, small
> (Collected by SB, JW, MM, 2002)
>
> 4 part sign reading: "KISER Rip Jew Town" (Rip is thought to be abbreviation for Rest in Peace)
> "Polish Sausage Pork Chops Hamburger Hot Dog Fries," 8′ × 20″
> "Free Fries with Sandwich," 8′ × 18″
> (Collected by SB, JW, MM, 2002)
> **MAXWELL STREET ARCHIVES AND ARTEFACT LIST**

The eleven-page, single-spaced list of items includes bits of buildings (tiles, frames, doors, ceiling fragments, etc.), signs, paper bags, bits of wood, ashtrays, books, papers, excavated soil samples, a lamp shaped like a bear, shirts, a sewing machine, records, pushcarts, and six tailor shop male mannequins. The list of objects reveals the interests of some of the people who collected the material. It was not simply buildings that interested them but material evidence of the life that went on in and around them. These are all undoubtedly noncontributing objects in the terminology of the National Park Service, but it is things such as these that made Maxwell Street what it was.

The failure of the Maxwell Street Historic Preservation Coalition to achieve historic-place status was based on the lack of "integrity" exhibited by the material landscape, which, as far as the National Register was concerned, had simply changed too much to effectively communicate the meanings and practices associated with the market and the history of the blues. The place no longer evoked the transitory nature of the activities that took place there. Ironically, though, it is exactly these links—between materiality, practice, and meaning—that are busily being reinscribed into the built environment. The transient qualities of the place have been appropriated by a university neighborhood renewal project that seeks to sell smart, modern apartments through a reminder of the place's history. The intersection of Maxwell and Halsted Streets now includes a collection of restaurants and shops linked by sidewalks featuring, as street furniture, bronze musicians and market stallholders playing blues and selling fruit. Practice and meaning are being re-encoded literally through material statues and façades accompanied by helpful notices reminding the visitor of the place as it once was. The developers of University Village seem confident that they can connect the materiality of this place to its past activities—a connection they deem strong enough to sell expensive apartments.

The endless play of change and persistence produces many ironies at Maxwell Street. UIC and the City of Chicago insist on the need to change the place from a flea market marked by waste and decay into a landscape marked by the aesthetics of neotraditional "new" urbanism—a movement that looks back to the past and insists on labelling the area a "village." Defenders of the market use the idea of preservation to protect it—linking the present landscape to a particular moment in the past. The National Park Service argues that the place cannot be saved from change as it has changed too much already, while the preservationists argue for the preservation of a lively place marked by constant change. One side wants to prevent the transformation of a place of change and liveliness, while the other want to change the place into a stable and economically valuable landscape—an "urban village." Each argument mobilizes different assemblages of materiality, meaning, and practice.

Archives

> We begin in the thick of things.
> CAITLIN DESILVEY[253]

A List of Listed Things

. . . poultry, crates, day-before-yesterday's spinach, fresh fish, garlic, cheese and sausage, prunes, gum-drops, rye bread, dill pickles and other kosher delicacies . . . shoes, clothing, fish, oranges, kettles, glassware, candy, jewelry, vegetables, crates of live poultry, hats, caps, pretzels, hot-dogs, ice cream cones, beads and beans, hardware and soft drinks, lipsticks and garlic,[254] clucking white pullets, geese, pigeons, rabbits and pet pups, new straw hats and vintage bird cages, musical instruments, fresh strawberries, crockery, ladies' hats in the latest cuckoo designs and kerosene lamps,[255] gypsy fortune-tellers, fake swamis, yodelling fish peddlers, and amateur entertainers, buyers, bargainers, and sight-seers,[256] silk dressing gowns, shoes, cotton dresses, jewelry, blankets, little tables, lemons, hot dogs, scrap iron, furs, superstitions, science, religion, food, physical and spiritual remedies, rags, silks, popcorn, guitars, radios,[257] glazed ceramic tiles in boxes; a telephone-answering machine; long-stemmed glasses and crystal goblets; gloves (ski and regular); automobile wheels and tires; underwear; jeans; jackets and jump suits; battery chargers; a snow blower; notebooks and paper for school; baseball trophies; skates (ice and roller); and comforters,[258] pungent odor of garlic, sizzling hot-dogs, spoiling fruit, aging cheese and pickled fish. . . .[259]

Writers are not the only ones confronted with arrays of things and asked to make some sense of them. Archivists have to do this in order to construct an archive—to use some principles of "appraisal" to winnow out the 1 to 5 percent of documents that will be archived for even our most important institutions.[260] And then scholars visiting the archive are confronted with the archivists' still-abundant selections, within which they in turn can experience "the deep satisfaction of finding things."

> You think: these people have left me the lot: each washboard and doormat purchased; saucepans, soup tureens, mirrors, newspapers, ounces of cinna-

mon and dozens of lemons; each ha'penny handed to a poor child, the minute agreement about how long it shall take a servant to get to keep the greatcoat you provide him with at the hiring; clothes pegs, fat hog meat, the exact expenditure on spirits in a year; the price of papering a room. . . . Everything. Not a purchase made, not a thing acquired that is not noted and recorded. You think: I could get to hate these people; and then: I can never do these people justice; and finally: I shall never get it done.

CAROLYN STEEDMAN[261]

Things are at the heart of the process of constructing an archive of a place and the existence of place *as* archive.[262]

What are the geographies of archives and archiving?[263] How has Maxwell Street been archived and how has it become an archive? How have the "things" of Maxwell Street become "archive things"? What kinds of value are placed on these things by various "gleaners"—people who collect and recycle things thought of as waste by others? "The archive" is not something that appears fully formed and pristine as if by magic but is created by many agents with often conflicting evaluations of what belongs and why.

Subject corp:	Maxwell Street Market (Chicago, Ill.)—Photographs.
Subj topic:	Markets—Illinois—Chicago.—1930–1949. lctgm
	Peddlers—Illinois—Chicago.—1930–1949. lctgm
Subject geo:	Chicago (Ill.)—1930–1949.
	Near West Side (Chicago, Ill.)—1930–1949.
Form/genre:	Gelatin silver prints.
	Photographic prints.

CHICAGO HISTORY MUSEUM, SPECIAL COLLECTIONS[264]

My consideration of the archiving of Maxwell Street is informed by those who urge us to give due care and attention to the things people push to one side and ignore, the things that do not make it into official places of memory.[265]

Engagement with the remaindered objects of a former mode of existence can activate a process of remembrance that carries threads of past lives into our present (and plies those threads up with our own imagined versions of those lives).

CAITLIN DESILVEY[266]

Caitlin DeSilvey evokes the archaeology/performance amalgam of Mike Pearson and Mike Shanks and their notion of a "rescue archaeology," which places emphasis on the high cultural stakes at play in linking seemingly worthless things to the "endless narratives, the political aspirations and disappointments, which have accumulated around them."[267]

How do people gather and attach value to things associated with a particular place—Maxwell Street? To answer this involves an account of the various practices of gleaning. It involved thinking of archiving as an active process.[268]

In the process of gathering things that are valued, an archive comes into being. This process of gathering, valuing, and archive-making reflects the abilities of those doing the collecting: their ability to give something value, ensure that this value is shared, and defend this value against countercharges of valuelessness or alternative values. Archives reflect these processes.

The "archive" means several things. For the researcher exploring history, the archive is a place to look for documents that will tell them about the past. For others, the archive is "a metaphoric invocation for any corpus of selective collections and the longings that the acquisitive quests for the primary, originary, and untouched entail."[269]

To some, thinking the archive is thinking of it as a sealed, special, imperial kind of place from which authenticity and history are judged. This view of the archive combines notions of power, durability, origins, place, and authority. In this sense, archival memory seems once and for all; it is carefully delineated, codified, and cataloged and is imperial in its scope.[270] The archive has been figured as a grid that disciplines its contents and erases undecidability.[271]

Historical Note

Maxwell Street is a famed street on Chicago's Near West Side. Crossing Halsted just south of Roosevelt, it developed as an open-market during the late 19th and early 20th centuries when Eastern-European Jewish immigrants populated the surrounding area. As the neighborhood changed during the Great Migration towards an African-American population, the Maxwell Street market became one of the best places to go to hear blues musicians. The cultural heritage of the neighborhood boasted such features as the filming site of the 1970's television show "Hill Street Blues," an early stage to musician and bandleader Benny Goodman, and the setting for the 1980 film *The Blues Brothers*.

In the 1990s, the University of Illinois at Chicago, located to the north of the Maxwell Street neighborhood, expanded its campus south and the Maxwell Street Market was relocated by the City of Chicago to Canal Street. As a result of negotiations with the Maxwell Street Historic Preservation Coalition, the façades of 13 buildings on Halsted, Maxwell and Roosevelt were relocated to new buildings on Maxwell Street and several existing buildings were adaptively re-used. The neighborhood today houses University student residence halls, academic buildings, parking facilities, retail establishments, and residential housing.

Scope and Contents Note

This collection offers pre-gentrification images of the original Maxwell Street before the relocation of the market and the demolition of most of the original

buildings. The collection includes twenty snapshot photos and negatives taken by Chicago-area author Gary Buslik in August 2000.
UNIVERSITY OF ILLINOIS CHICAGO, SPECIAL COLLECTIONS[272]

The word "archive" comes from the Greek ἀρχή (*arkhē*), which means "magistracy" or "government." The same root can be found, with a negating prefix, in "anarchy."

Archives are far more contingent, messy, and permeable than notions of the sealed, imperial archive suggest. Some have raised the possibility of alternative archives as non-sanctioned places of excluded or marginalized memory.[273] But even official archives are sites where it is possible to read against the grain and find unofficial stories in the absences and unintended presences.[274]

While the contents of the archive are important, they need to be understood hand in hand with the mechanisms through which the archive is produced and its contents made public. Reading archives this way—along the grain—allows us to delineate how power makes up its claims to truth within the construction process of a particular archive—how the process of selection and categorization produces its own historical ontologies, its own possibilities for what exists and what is true.[275]

Place is an archive too.

> People interpret material traces to reconstruct past events, the conditions under which such interpretation takes place, and the role this interpretation plays in historical consciousness and social memory. . . . As different groups struggled to control the fate of a region and its resources, they invoked very different understandings of its past to justify their actions. In many cases, their beliefs about the region's history were informed, directly or indirectly, by physical evidence found in the place itself.
> **WILLIAM H. TURKEL**[276]

Nate's Delicatessen was a local landmark on Maxwell Street, and by the 1990s it was the only restaurant where you could sit inside at a table. Nate prepared kosher corned beef and pickled herring from recipes of the deli's former Jewish owners, from whom he bought the business as their former employee. Members of the Maxwell Street Foundation salvaged Nate's Delicatessen marquee when the building was demolished in 1995, and temporarily stored it in a nearby building with permission.
MAXWELL STREET FOUNDATION[277]

Various forms of expert knowledge are engaged in documenting the natural and human environment of a place. These expert knowledges are connected to the presence of particular kinds of *things* in the landscape that, with their knowledge, can act as an in-situ archive containing a number of different kinds of memory.

One way to account for the process of archive construction is through a kind of archive ethnography:

> By ethnographies, I simply mean descriptions of historical and contemporary cases that focus closely and thematically on the work of assembling, disrupting and reconfiguring particular archival collections. . . . For such ethnographies, archives would be salient less as methodological resources for historical studies than as historical phenomena in their own right. Such an orientation does not negate the scholarly use of archival information; instead, it shifts attention to archives in formation and the localized gathering of histories.
> **MICHAEL LYNCH**[278]

These "archive ethnographies" serve to show how the passage of documents and artifacts from private to public precedes the institutionalization of an archive—"and this passage can be a site of struggle, occasionally resulting in breach, abortion, or miscarriage of the nascent archive. Consequently, we can appreciate that archives are as much products of historical struggle as they are primary sources for writing histories."[279]

Value

The question of what gets included in an archive is a question about value. Few now believe, following Kant, that value is a quality inherent to an object. Value emerges contextually and relates to the interests of those doing the valuing. Indeed, seeing value as intrinsic to an object only hides the interests of those doing the valuing, and the closer we get to the top of social hierarchies the more likely it is that valuation includes some notion of pure and unpolluted value in an object.[280] Science and connoisseurship both partake of this illusion.

Disinterestedness hides particular interests.

> There are no statements of value, or evaluations, which are not statements of particular needs, desires or preferences, whether of individuals, or of groups.
> **STEVEN CONNOR**[281]

There is, undoubtedly, a bourgeois tint to the fascination with flea markets. The bargains available there represent "the choice of the necessary" to the poor who have no choice but to shop there. To those like myself, who enjoy the aesthetics of waste and detritus, there is an aestheticization of the world that those who are forced to inhabit it would rather do without.

> "Those are the hands of someone who has worked too much, doing very hard manual work. . . . As a matter of fact it's very unusual to see hands like that" (engineer, Paris). "These two hands unquestionably evoke a poor and unhappy old age" (teacher, provinces). An aestheticizing reference to paint-

> ing, sculpture or literature, more frequent, more varied and more subtly handled, resorts to the neutralizing and distancing which bourgeois discourse about the social world requires and performs. "I find this is a very beautiful photograph. It's the very symbol of toil. It puts me in mind of Flaubert's old servant-woman. . . . That woman's gesture, at once very humble. . . . It's terrible that work and poverty are so deforming" (engineer, Paris).
> **PIERRE BOURDIEU**[282]

Sensing Maxwell Street Market in terms of surrealism and pattern is a bourgeois distancing from the lived experience of many who lived around the market.

Objects pass through "regimes of value." Rubbish and waste can shift between valuelessness and value where the decay over time of an object does not necessarily lead to a loss of value.

> Signs of aging and use can contribute to increased or auratic value. In the same sense the shiny new plastic cup appears to us as imminent rubbish; disposability makes transient value strikingly visible. The fact of malleability and transformation in value is evidence that objects are not locked into categories because of their material qualities. It is how their materiality is apprehended and used that is the key to value and its transformation.
> **GAY HAWKINS**[283]

One of the axes around which value is inscribed is the axis of transience and durability, where durability is valued and transience is not. In his wonderfully titled *Rubbish Theory*, Michael Thompson insists that the key liminal category between the durable world and the transient world is "rubbish." Rubbish provides a categorical space where transience can be converted into durability. Rubbish denotes objects that have the possibility of being revalued as durable—as valuable. One of Thompson's examples is run-down inner-city houses that are gentrified and revalued as "desirable." The assignment of things to the category of durable or transient is performed by those in control of time, space, and knowledge, who thus ensure "that their own objects are always durable and that those of others are always transient."[284]

Value is produced by the passage of things in and out of different *regimes of value*—

> those manifolds in which judgements of taste crystallize and take hold: conflictual apparatuses which decree that in one context and for one social group a particular house is a derelict slum, and in another context and for another social group is a stylish worker's cottage with infinite potential for restoration.
> **JOHN FROW**[285]

Objects have biographies that are formed as they pass through these regimes of value. The same object may be seen as valueless and thus mere waste in one regime and as precious and worthy of being granted durability in another. The value of an object is thus not defined by any sense of its essential properties or functional utility.

> Every humanly made object, no matter how grand or trivial—and regardless of what happens to it as some later point in its life history—starts out as something "genuine," something possessing at least a minimal degree of worth. Every object, for example, has its own form, its own aesthetic, its own integrity, and its own identity, and every one contains some amount of human expressivity, even if mediated by machine. . . . Nevertheless, every genuine object that enters the world is subject to the same natural depredations, the same ravages of time, that befall all material things. Objects get used or used up, worn or worn out. Usually their fate is to move steadily and inexorably from the status of objects to the status of waste unless or until something intervenes to slow or even reverse this process, in which case the object does not become junk but rather just the opposite: it retains or even in increases in value.
>
> **DAVID GROSS**[286]

Things have an intimate relationship to memory. We remember events and happenings, but it is often things—artifacts—that trigger our memories. Some things endure through time and carry memories with them.

> Artifacts are thrust into the world. They have the power to stabilize life. Transient feelings and thoughts gain permanence and objectivity in things—in the jugs and chairs that endure.
>
> **YI-FU TUAN**[287]

What intervenes in the biography of an object that makes it valued and valuable? One answer is the work of archiving.

The archive is a particular kind of place where objects are valued, with its own regimes of value.

> Differences abound on the principles of concepts (or theories) that should animate appraisal or that define the "value" or "significance" or "importance" of records—all terms used by archival legislation and by archivists, usually without definition or reflection. "What makes the good?" the Greeks long ago asked. What makes something have value, be worth preserving and remembering? Not surprisingly, without clear first principles, the resulting strategies and methodologies have achieved no consensus.
>
> **TERRY COOK**[288]

Often archives revolve around regimes of value that rest on a notion of disinterestedness that privileges notions such as aura, authenticity, and origins. Archivists are supposed to use the process of appraisal dispassionately, denying the active role of interpretation in the process of selection. In fact, archivists are valuing things in a way that not only constructs the past but preserves the future.

> Do you know this wedding photo?
>
> The Maxwell Street Foundation maintains a collection of items since 1995, some of which were delivered to our hands anonymously having come from former Maxwell Street neighborhood buildings and storefronts. This wedding photo is an example of that, and it is posted here in the event a visitor to our website may recognize people in this photo. It was taken at Ridge Studio here in Chicago, 639 W. 120th St., and we believe it is circa 1950s. We want to know the Maxwell Street story behind this photo!
>
> **MAXWELL STREET FOUNDATION**[289]

For an archive to exist, things (paper, documents, objects) have to be collected. The act of collecting is part of the act of valuing. From an infinite array of things, people choose an infinitely small selection and by doing so inscribe the chosen things with value. Sometimes that value may be entirely personal, and at other times (if the person is an archivist, for instance) a regime of value may be shared. People who collect are susceptible to the charms of things—of some things and not others.

> What is decisive in collecting is that the object is detached from all its original functions in order to enter into the closest conceivable relation to things of the same kind. This relation is the diametric opposite of any utility, and falls into the peculiar category of completeness. What is this "completeness"? It is a grand attempt to overcome the wholly irrational character of the object's mere presence at hand through is integration into a new, expressly devised historical system: the collection.
>
> **WALTER BENJAMIN**[290]

Benjamin's account of collecting celebrates the private and chaotic collection (particularly his own books in the process of unpacking them) at the expense of the formal, carefully ordered and cataloged, public collection of the museum or library. Collecting represents a personal reaction to the enchantment of objects.

Gleaning, Gleaners

In François Millet's painting *Les Glaneuses* (*The Gleaners*), three women collect the leftovers of the harvest, a traditional practice in nineteenth-century France among the rural poor, who were granted the right to pick over the remains of the production process. In 1857, when it was first exhibited, *Les Glaneuses* was met with hostility by some critics, who saw the women as grotesque and as an unwelcome reminder of the marginalized poor. They did not conform to notions of appropriate beauty that existed at the time.

This artistic representation of gleaners was referred to and updated in Agnes Varda's film *The Gleaners and I* (2000), in which she illuminates the lives of contemporary waste collectors in France, such as those who, in a modern

FIGURE 48.
The Gleaners, *1857. Painting by Jean-François Millet.*

evocation of Millet's painting, collect potatoes that are not the right shape for sale in the formal economy. Her film shows us how both waste and the people associated with waste are the products of particular forms of ordering. The potatoes gleaned by contemporary French gleaners are waste only because of the demands for particular sizes and shapes of potato imposed by supermarkets. Potatoes have to conform to some notion of potatoes. Anything less or more and they are not-quite-potatoes and can be left in the fields or simply dumped as waste: left to the gleaners. Varda's philosophical film reconfigures the traditional subject—the gleaner—into anyone who values waste, decay, and ephemerality. A new species of flaneur perhaps—with a waste specialty.

> Gleaning involves taking what is discarded and re-envisioning its worth. By retrieving discarded objects and placing them in relation to each other new associations and meanings are released.
> **GERALDINE PRATT**[291]

Gleaning, then, produces a kind of inverse archive, much like the one envisaged by Bataille, which "included that which is excluded, devalued, expelled or effaced" and involved a "magic reworking of the archive through taxonomic disorder."[292] In a neat turn, Varda's own practices of filming are shown as a kind of gleaning as she captures images of herself during a road trip. In this sense, the journalist writing about Maxwell Street is as much a gleaner as the people he is writing about. I too have been a gleaner of information about Maxwell Street for much of the past decade, as I have eaten tamales at the market, strolled the streets of the area, and talked to the various gleaners of Maxwell Street things I have encountered.

Collecting is not gleaning. While gleaning, like collecting, involves a kind of valuing, it operates in different regimes of value. Varda's gleaners value the misshapen potatoes that are deemed lacking in value by the supermarkets. The formal harvest is a kind of collecting distinct from gleaning. Similarly the work of the formal archives and of the development companies at Maxwell Street involved a process of formal assessment and valuing that cannot easily be described as gleaning. My focus is on those who tried to value what was left over and what was destroyed in the process.

One of the bronze statues that now decorate the streetscape on the corner of Maxwell and Halsted is of a pile of artfully "untidy" wooden crates signifying the kind of untidy and disordered place this once was. In its hundred-year history, Maxwell Street Market was a world of promiscuous "things" tumbling out of the buildings that lined the street, clogging the spaces in between, and inviting the attention of those who came to experience the buzz of the market.

> My early methods were more akin to gleaning than to systematic inventory, and I often wasn't sure what to do with the things I unearthed.
> **CAITLIN DESILVEY**[293]

Gleaning, here, is figured as an informal, unorganized form of gathering stuff. I want to hold on to the notion of gleaning, as it befits the kind of place Maxwell Street was through the late nineteenth and twentieth centuries.

The Chicago History Museum is an obvious place to explore the histories of Chicago streets and neighborhoods. It has a series of clippings files labeled by street, park, neighborhood, and so on. There is a large manila folder labeled "Maxwell Street" with cuttings dating from over a century.

> In all the wildernesses of narrow by-streets, garbage boxes are everywhere. They are not the boxes, greasy and foul smelling, which are so often at the backdoors of Chicago's mansions on the avenues; they line the sidewalks in front of the houses, mile in and mile out, encroaching on the space for pedestrians and worn shiny and smooth from the lolling contact and street gossips. As gleaners from the garbage boxes of every other section of this great city, the wonder of it is that these people should have such boxes at all. Their lives and interests are so based on the extravagant habits of others that it would seem that they should have nothing to throw away.
> ***CHICAGO TRIBUNE*, 1896**[294]

FIGURE 49.
Hubcaps, Union Street, north of Maxwell Street, *1966. Photo by James Newberry. ICHi-20332. Chicago History Museum.*

There was long a suspicion that many of the things to be found at Maxwell Street were stolen. In some cases the origins of the products were clearer. In 1974 one reporter discovered seventy-two-year-old Leamon Reynolds next to a six-foot-high pile of hubcaps selling for around $1.50 each. Reynolds "can find you a 1952 Ford hubcap in five seconds, thanks to his secret filing system. Where does Reynolds get his fantastic stock? From state road workers, he says, they pick them up while patrolling expressways."[295]

Six years earlier, Reynolds's hubcap stand had attracted the attention of the photographer James Newberry, who was entranced enough with its silvery stock to take a picture that can be found today in the archives of the Chicago History Museum. The picture can also be seen on the remaining block of Maxwell Street, on the side of one of the mock-piles of boxes that remind the present-day visitor of the place this once was. In this way, the humble hubcap found its way first into the official archive of the market and then back, in simulated form, onto the street where the market once was.

As with journalists addressing chaos through list-making, photographers were entranced by both the weight of objects and the aesthetics of juxtaposition. Newberry's image of hubcaps and brooms is like a still life. There are no people. Many of the photographs in the archive are of churning crowds of peoples and varieties of performance that take place in a market. This, on the other hand, appears as an accidental arrangement of confused forms and

FIGURE 50.
Hubcaps on mock pile of crates sculpture on Maxwell Street, 2009. Photo by author.

surfaces. The beauty of the photograph lies in the configuration of the disks of the hubcaps and the straight lines of the brooms. It presents us with aesthetically pleasing confusion.

Things come together and appear in a flea market in ways that would not happen in a carefully organized department store. This is the magic that is replicated in the words of journalists who replace the (dis)organization of objects with words on a page. "Musical instruments, fresh strawberries, crockery, ladies' hats in the latest cuckoo designs and kerosene lamps."

While it failed to prevent the removal of the market from its original site, the Maxwell Street Historic Preservation Coalition was able to collect fragments of the site as it was progressively demolished. Its members scavenged and gleaned among the abandoned buildings of the area, and after demolitions among the rubble. They knew that the place was ill-fated and that they could not preserve the market. But something of the spirit of the place survives in its things. In an appropriate setting they might retain some of their site-specificity, and this is what the gleaners were counting on.

As the area was progressively demolished, members of the coalition collected an unlikely list of things that were carefully cataloged and stored in back gardens, churchyards, and basements around Chicago.

Archives/ Things

I was introduced to this archive of things from the Maxwell Street area by Lori Grove, the author of a beautifully illustrated book about the area and a key player in the Maxwell Street Historic Preservation Coalition.[296] Grove is soft-spoken and meticulous. Her love of the market's things extends to art-deco details and pieces of wood that may have been in the area as the Civil War ended—before the Great Fire and Chicago's dramatic growth. Her record of the objects on these shelves is clear and easy to understand. Her files of papers relating to the efforts of the coalition are carefully organized and indexed. She is swimming in history. She knows her architects and their styles. She can tell me about almost any address in the vicinity of the market. She wants the objects in the collection to find a home. All the protecting, ordering, and cataloging is done with this in mind—to produce an archive that might be recognized as an archive. The efforts to save the Maxwell Street area have (mostly) failed. Grove's efforts now are directed toward archiving. Grove works in the museum sector. She is a professional. In the past she has worked as a scientific illustrator—a skill that requires fastidious attention to detail in addition to a keen aesthetic sensibility. This is evident in her careful management of the objects in the collection.

Most of these objects have now been moved to a storage facility generously donated by a local business. They are housed on metal shelves in a small, locked area of a warehouse and metalworking shop. They are arranged according to where they came from, with larger objects stored wherever there is room. You can find the things listed in the careful inventory of the archive and its artifacts. You can find, for instance, a list of things gleaned from 719 West Maxwell Street, once home to "Paul and Bill the Tailors" shop.

> 1 treadle sewing machine installed in wooden cabinet, w/work light (from the 1st floor, south)
> 1 "edger?" machine installed in wooden cabinet w/work light (from the 1st floor, south)
> 1 single sewing machine (from the 1st floor, south)
> 1 work table (from 2nd flr., north)
> 1 store display table (from 2nd flr., north)
> 1 set of wooden horses (from 2nd flr, north)
> 1 wooden door (that was set on the horses as a table-top, from 2nd flr., north)
> Approx. 20 pair of unfinished men's slacks (1st flr., north) . . .
> **MAXWELL STREET ARCHIVES AND ARTEFACT LIST**

Grove shows me the ancient-looking sewing machine rescued from the remains of 719 West Maxwell Street. The tailors shop was the last business on the old street to close, and it seems special to Grove. She continues to show me things. One box contains a frilled white shirt still wrapped in cellophane. On another shelf there is a 1920s wooden mold for making a boater hat. An empty "trade accounts" receipt book; a box of dusty glass jars, one with the words "Three generations of babies have loved Bowman's Milk" (from

1225 South Halsted); a sale tag that reads, "Now only $1.96"; old buttons and bits of stone; spools of thread and rubber stamps; whole window frames and larger bits of decaying wood; a faded landscape painting that once hung on the wall of a shop. On some shelves, carefully placed in cardboard trays, are bits of splintered building wood cracked along the grain where nails used to hold some structure together. At the end of the storage area are things too big to fit in boxes or on shelves: a green sign, its metal rusting, for Gethsemane Missionary Baptist Church; a Maxwell Street street sign; sections of pressed tin-plate roofing; bits and pieces of buildings. One piece of wood, from 716 West Maxwell Street, may date from the Civil War. Grove hopes to get it scientifically dated.

Things from Maxwell Street have also found their way to other places in the city. Some particularly large pieces of architectural material and a large metal sign are stored in the basement of a condominium complex to the north of the Loop. I was shown a locked room, among smaller spaces where residents stored their bikes. In it were sizable chunks of the old landscape: a large section of splintered wood; a fragment of large green, metal lettering from a sign; chunks of ornamental concrete; sections of tiled floor on pallets; most of a Maxwell Street seller's cart. A yellow sign on the wall reads

What is all this Stuff?
These are architectural artefacts salvaged from
buildings in the Maxwell Street neighborhood.
They are the property of the Maxwell Street
Preservation Coalition. The xxxxxx
Board has given me permission to store them
here. Please do not disturb them.

There is certainly something poignant about all this stuff. Old labels, signs, shirts, rotting pieces of wood. Without labels or explanations for why they have been deemed worthy of gleaning there is a sense of a lived place. The contents of the emergent archives on Maxwell Street suggest the bustle of trade and the entropy of a derelict urban landscape. Sale tags and blank receipts bring to mind the constant competition at the low end of a retail geography. A shirt indexes the exaggerated glamour of impoverished dandies. A sewing machine links the new gleaners to the sweatshop-like labor of the Jewish clothiers who once pulled in passersby with noisy performative gusto. Old pieces of wood with flaking paint and nail holes, or a piece of green awning, point toward a material landscape that played a secondary role to the business of the market performance.

If there is one principle that informs the collecting of things for this archive it is the connection between objects and the *activities* of the market. Performance theorist Diana Taylor has suggested a cleavage between the archive (as a formal material repository of knowledge) and the repertoire (a practiced

FIGURE 51.
Sewing machine from 719 West Maxwell Street. Photo by author.

and reiterated transmission of knowledge through performance).[297] In this archive it is precisely the connection between things (sales tickets, shirts, sewing machines) and the repertoire of a flea market that is at stake.

But even among the collections of the Maxwell Street Historic Preservation Coalition (which are all outside of the formal spaces of memory), there is a hierarchy of what counts and what does not. In many ways the most telling collection of all can be found in the house of Steve Balkin, one of the more outspoken protesters from the coalition. Over three stories, from bottom to top, the house he shares with his wife is an archive of Maxwell Street things. At the head of the "Maxwell Street Archives and Artefact List, 3/04" is a list of sites where artifacts were held before more appropriate storage could be found. One line reads, "SB = Steve Balkin's house and office (complete collection not included: TBD)." The collection there is immense. There are things from the old market. There are documents relating to the struggle to save it. Balkin continues to shop at the new, relocated Maxwell Street Market, and he delights in kitsch. Most of the things in his house are objects that have not become part of the listed material archive of the coalition.

Balkin's house resembles the market in its chaotic clamor of flyers, statuettes, faded photos on the walls, boxes of papers under tables. Maxwell Street is everywhere. This is about as far from a formal archive as you could expect. This is not the "first place"—the site of strategic authority that reaches out over space and time with imperial vigor. This is knickknacks and unwanted things, garish colors, plastic, torn pieces of paper, memorials to hot dog stands, homemade placards once carried in protest while the ramshackle space of Maxwell Street was being erased and replaced. It is a space that feels like it excludes nothing, embraces everything. Maxwell Street is alive in a north-side house. It would not be possible to read a narrative of Maxwell Street from the arrangement of things here—at least not a linear one. The collection has not been cataloged or ordered. It reflects the magpie tendencies of Balkin and his intense love of all things connected to the market. Balkin is a professor of economics at Roosevelt University, who works on microeconomics and open-air markets. He is a bundle of energy. He is frequently at the market and has been for decades. He films and photographs it endlessly. Once we had met I started to receive a constant stream of emails referring to the injustices surrounding informal market economies around the world. He must have written thousands of letters to newspapers and government officials about Maxwell Street and its plight. There is no question that to Balkin the issues surrounding the area concern social justice, racism, gentrification, and urban purification. His stance toward this place is both radical and loving.

Under a table in an upstairs kitchen room are boxes and boxes of papers relating to the Maxwell Street Market and the coalition's attempts to save it. They are not ordered or indexed in any apparent way. They include flyers, posters, signs used in protests, copies of letters written to newspapers and government officials, notifications of blues concerts, seemingly endless scraps of paper, often in duplicate or triplicate. As he rummages through these boxes he hands me things to keep, copies of things he already has. He hopes I can use them.

During two tours of Balkin's home he mentions his desire to get rid of some of this stuff. He wishes it could be properly archived alongside all of the other objects the coalition has collected. He muses over the possibility of a student being given access to all this material in return for cataloging it. He suspects that there is little use for much of it. To Balkin, this extraordinary collection is a direct connection to years of protesting and advocating on behalf of the market he loved.

On the walls of Balkin's house are old postcards and prints of the market. On his shelves are objects he has bought at the old and new markets: a lamp that appears to be a bald man attached to a fire hydrant, many kitsch statuettes, a brightly colored oval print of a hobo-clown, angels playing accordions, a glowing red furry Valentine's heart. This is a very different collection of things. It is kitsch archive.[298]

The top floor of the house is Balkin's office. Like the lower floors it is full of Maxwell Street stuff—pictures, T-shirts, and baseball caps from the protests, ornaments from the market, and the vast collection of film he shot as the market was being slowly erased.

FIGURE 52.
Boxes of Maxwell Street material under a kitchen table in Steve Balkin's home. Photo by author.

> The hoarder's basic fantasy is the fantasy of *reuse*. It is his hope that the lost use value of an empty plastic bottle or an old magazine will one day, in the unknown future, be restored. An object that the good consumer considered dead is kept by the hoarder in anticipation of what may be called its materialistic resurrection.
>
> **DAVID KISHIK**[299]

Far from being insulated from alterity, archives are porous places. The Maxwell Street archive is distributed across multiple sites that bleed into each other. I have spent many hours in the Chicago History Museum investigating all of the material filed under "Maxwell Street" (and related headings). Other relevant archives exist at the Harold Washington Library, the University of Chicago, and the University of Illinois at Chicago. At UIC the library's special collections department is situated more or less on top of the site of the old market. These are the obvious archives for anyone exploring the history of the market, and I have dutifully done my time. But there are other archives. There is 41°51'53" N, 87°38'49" W—the corner of Maxwell and Halsted, where the market used to be at its liveliest. There is the relocated market on nearby Desplaines Street. These places are also archives. Then there are the coalition's storage facilities and Steve Balkin's house.

And things move between these archives. Along the existing 700 block of Maxwell Street, where it intersects South Halsted, are the thirteen façades of buildings that were moved from around the area to be joined together and placed in front of parking lots. These façades represent a victory of sorts for the Maxwell Street Historic Preservation Coalition, as they were saved from complete demolition. These things were considered officially valuable due to architectural details that had survived over the years. The ground floors house various shops—mostly franchises of small chains. On the street are the bronze statues and plaques representing the street's history. These include images from the archives in the Chicago History Museum. Inside a university residence block is a wall decorated with a mural-size reproduction of a photograph of the street at its busiest. Inside the window of an empty store are prints of images by various photographers (including Nathan Lerner) that are being sold to raise money for the coalition. This place is an archive that "bears material traces of the past in the very substance of the place."[300]

As you enter the Chicago History Museum, you will see signs for Jim's hot dog stand, salvaged from Maxwell Street, proudly displayed on the lobby wall. In the research room, in addition to the newspaper clippings and photograph files, you will find the following entry in the catalog:

Title:	Maxwell Street film footage [videorecording]
Author/Creator:	Balkin, Steven
Pub./made:	1992–2000
Phys. desc.:	45 DVDs. (Copies). 45 Hi8 videotapes. (Originals).
Summary:	Film footage showing vendors at their stands, shoppers and shopkeepers, visitors, and street musicians, and the landscape of the Maxwell Street Neighborhood. . . .

Balkin shot hundreds of hours of footage of the market as it was being erased from the landscape. He took thousands of photographs too, which are also available at the Chicago History Museum. The archives flow into each other.

Kitsch

Think about the difference between objects picked out for being either unique or representative and placed in a special place (the "archive") and the kitsch that characterizes many of the objects in Balkin's house. Kitsch has been defined as an aesthetic marked by its repetitive, mass-produced qualities, one that "employs the thematics of repetition over innovation, a preference for formulae and conventions over originality and experiment, an appeal to sentimental affirmation over existential probing—a unique and quite 'healthy' sensibility."[301] Kitsch has the qualities of "eclectic cannibalism, recycling, rejoicing in surface or allegorical value," which contrasts with the more serious belief in "authenticity, originality, and symbolic death."[302]

Such evenhanded accounts see repetition not as inferior to the new and creative but as equally valid and powerful. The power of kitsch lies in the fact that it is not distanced from everyday life, in a gallery or official archive, but embedded in the mundane—an embeddedness that shores up life against the uncertainties of shock and change and innovation:

> It nestles right into the mundane, savoring its secure patterns and its meter. This is also why kitsch appears so often in the real spaces of the quotidian itself (offices, waiting rooms), in the decorative, the comforting and the trivial trimmings of daily life.
> **SAM BINKLEY**[303]

Insofar as kitsch relies of repetition rather than uniqueness, it has no place in the archive. By collecting objects, classifying them, and storing them, the coalition members are attempting to produce an alternative archive and to save something of their memory of the market, affirming value in objects that if not preserved and archived would be simply a form of waste. The chaotic and unordered contents of Steve Balkin's house are still further from what counts as would generally be deemed valuable objects. Here we see a hierarchy of value and a variety of what counts as an archive of the Maxwell Street area. Insofar as objects do not count as valuable enough to be preserved and enter the archive they are simply a form of waste. They have no "aura" and appear to be disposable and replaceable.

Maxwell Street Market was always a space apart from those designated as proper places for objects of high value. It was a space where things gleaned were repackaged and resold as bargains. There is an irony in making bits of the market (whether cellophane-wrapped shirts or pieces of old awning) special in the way that archives normally work. This was a place that sold things associated with the marginalized in urban space—the immigrants and the poor. The things they wanted to buy, at cut-price rates, things necessary for everyday life. The "things" of the market were suffused with the politics of class and race and the links between consumption and petty crime. These are things that have come to have cultural value only because of the work of people like Grove and Balkin, who labor to make them significant.

Endings

These days, you're unlikely to be struck by any particular smells at the intersection of Halsted and Maxwell. The sounds are similarly unremarkable. There are no live chickens for sale, and nobody wears a chicken on their head. The transplanted façades that line the 700 block of Maxwell Street are attractive enough—cleaned, restored, and moved.

There are, nonetheless, similarities between present-day Maxwell Street and the street of the past. It is still made up of things from elsewhere. The crates of squawking chickens have been replaced by shipments of chicken wings, delivered to WOW Wingery at 717 Maxwell Street to be served with its array of "ethnic" sauces. People still buy things here, and the shops still get their stock from elsewhere. People live in the neighborhood and use the street. I have seen children climbing on the bronze replica packing crates.

The Maxwell Street of old has been described as a "third place"—a "great good place" beyond home and work where strangers meet and interact.[304] Such places, we are told, are disappearing. Place has been replaced with placelessness, authenticity with inauthenticity.

> An authentic attitude to place is thus understood to be a direct and genuine experience of the entire complex of the identity of places—not mediated and distorted through a series of quite arbitrary social and intellectual fashions about how that experience should be, nor following stereotyped conventions. It comes from a full awareness of places for what they are as products of man's intentions and the meaningful settings for human activities, or from a profound and unselfconscious identity with place.
> **EDWARD RELPH**[305]

> Authenticity differentiates a person, a product, or a group from its competitors; it confers an aura of moral superiority, a strategic advantage that each can use to its own benefit. In reality, few groups can be authentic in the contradictory ways we use the term: on the one hand, being primal, historically first or true to a traditional vision, and on the other hand, being unique, historically new, innovative, and creative. In modern times, though, it may not be necessary for a group to *be* authentic; it may be enough to claim to see authenticity in order to control its advantages.
> **SHARON ZUKIN**[306]

In the case of Maxwell Street, it is tempting to say that an authentic place has been stripped of its authenticity. I don't find this helpful. The people who live in University Village live lives that are every bit as authentic as anyone else's—going to work, raising families, or not. Places change, and this place too will one day be a place of the past that some will look back on with longing and nostalgia. At least some of the people who live here buy into a romantic vision of the market in its heyday. The apartments and town houses were sold as "yesterday's heritage." "University Village" became "University Village Maxwell Street" in 2015. The connection of the currently existing locale to its past was considered too weak and lacking in "integrity" to designate it as a "historic place" and save it from demolition, yet the same connection is clearly strong enough to sell high-end homes.

There are only two contiguous blocks of Maxwell Street remaining. Despite this, Maxwell Street is a distributed place. Pieces of it are in basements and backyards across Chicago thanks to the efforts of the Maxwell Street Foundation. Maxwell Street still appears in the form of Maxwell Street Days in Madison, Wisconsin. You can buy, or download, Jimmie Lee Robinson's *Maxwell Street Blues* CD. You can watch documentaries about the market or admire

Vivian Maier's increasingly lauded photographs. You can read Willard Motley's novels. You can find it in the catalogs and file drawers of the Chicago History Museum.

The question "Where is Maxwell Street?" is only partially answered by the coordinates 41°51'53" N, 87°38'49" W.

I would guess that the majority of people who now visit Maxwell Street do not actually get to Maxwell Street. If you look Maxwell Street up on Chicago tourism websites you will be directed to the relocated Maxwell Street Market.

> GETTING THERE
>
> Maxwell Street Market is located at 800 S. Desplaines St. . . .
>
> The Maxwell Street Market is a Chicago tradition of bargains and bargaining with an international flavor. The market offers an eclectic mix of merchandise—from tools to tires, plus fresh produce, furniture, clothing, rare finds and collectibles—and some of the best Mexican and Latin street food in Chicago.
>
> **CITY OF CHICAGO WEBSITE**[307]

To find Maxwell Street go to Desplaines Street.

I visited both Maxwell Street and the new market on my last visit, in 2012—the hundredth anniversary of the official recognition of the market. The market was lively. A clown made balloon animals for children. As usual I took photos of things. I bought some tacos and an horchata. A young man, I would guess in his early thirties, was watching the market wide-eyed. Recognizing me as the only other white face at the long wooden table, he engaged me in conversation. He lived in University Village and this was his first visit to the market. "This is so cool," he said.

In August 2000 the *Chicago Tribune* found Al and Mary Geiser admiring their new home in University Village. When asked what drew them to the development, they replied, "We liked the multi-ethnic feeling and the authenticity of the neighborhood—we're intrigued by the history of Maxwell Street."

PART THREE

Thinking Place

> . . . that without which nothing else can exist, while it can exist without the others, must needs be first.
>
> **ARISTOTLE**[1]

How do we theorize place in a way that takes us beyond the opposition between, on the one hand, confining, bounded, "reactionary" senses of place that focus on rootedness, attachment, and singularity, and, on the other hand, a distributed, open, "progressive" sense of place that focus on flows, connections, and networks?

Too often we make claims for new theoretical approaches by either ignoring or disparaging older traditions. Here I want to insist on the productive continuities and overlaps between the new and old.

In the first part of this book, I contemplated the problem of how to write place. In the second, I meandered through Maxwell Street picking up things that caught my attention. In this part, I use these fragments and snippets of Maxwell Street to construct a theory of place that might, at least in part, be applicable elsewhere, in other places. Like the bargains bought at a flea market, these trinkets find a new home, in a new assemblage.

If part 2 was an exercise in *local theory*, where conceptualization stayed close to the market, part 3 is *mesotheory*. Mesotheory is a little less modest than local theory. It reaches out to other places and provides a framework that may be useful for explorations elsewhere. Some may ask what grand theory, critical theory, or social theory this is, in turn, informed by? I cannot help except to say that this outline of a theory of place may be of use to those who wish to construct theory from a number of critical perspectives, from phenomenology to feminism, from Marxism to postcolonialism. It is informed by all of these but reducible to none of them. I offer it as a set of modest tools.

The modesty, however, is limited. I insist there is something fundamental about place as a *necessary social construct*—something that necessitates understanding regardless of the particular body of social theory that we want to engage or promote.

Suggesting a theory of place is of course dangerous. Suggesting a theory of anything is dangerous. There are two particular approaches to place that we need to tackle and transcend. The first is a broadly *phenomenological* approach that insists on the immediacy of place as experienced in an unmediated way. Such an approach depends on the bracketing off of the conditions for the production of any experience of immediacy. It asks, essentially, how place is experienced and what the role is of place in human experience. The second is broadly *materialist* or, to follow Bourdieu, *objectivist*.[2] This approach to place insists on prioritizing the context for experience—the "objective relations"—as a way of explaining exactly how and what we experience. It asks how a particular experience is possible. This is most clearly the approach of Marxists. What follows is neither of these and both of these.

A Necessary Social Construct

Some theorists of place, inspired by phenomenology and, particularly, Martin Heidegger, see place as an essential aspect of being-in-the-world, recognizing, with Aristotle, that for a thing to exist, it has to be some*where*. More radically inclined theorists insist that place is socially produced and the job of the theorist is to diagnose how, why, and to what ends. Perhaps place is both of these things at once—a *necessary social construct*.

A necessary social construct is something that in each instance is socially produced but, nevertheless, we cannot imagine being without. The figure of the refugee is a social construct. Without nation-states and all the documents and declarations that define what it is to be refugee, refugees would not exist. Indeed, there was a time when there was no such thing as a refugee. Place, on the other hand, has existed as long as humans have existed, and it is not possible to imagine a more-than-human world without place.

It is possible (for me) to imagine worlds without gender or class, capitalism or patriarchy. It is not possible (for me) to imagine a human world without place. This is an important difference.

Necessary social constructs, including place, are particularly powerful aspects of existence. Their appearance as necessary, as natural, makes them powerful elements in the production and reproduction of the social.

> One sentence, several sentences, a paragraph, can't sum up Maxwell Street. Its rowdiness defies a terse statement. Its lewdness escapes a brief summary. Maxwell has been called "Jewtown," the "ghetto," the "melting pot." It is all this and much more. Everything that is low and foul can be found there. And much that is fine and noble, sympathetic and generous. Clean

youths and pretty girls. Innocent kids playing tag and "red-light." Genteel old people.
WILLARD MOTLEY[3]

Location

"Place" can be thought of as a segment of space that has accumulated particular meanings at the level of the individual and the social. In the classic definition provided by Yi-Fu Tuan, place, as a field of care and center of meaning, is a pause in a wider, more abstract field of action.

> Space is a common symbol of freedom in the Western world. Space lies open; it suggests the future and invites action. On the negative side, space and freedom are a threat. A root meaning of the word "bad" is "open." To be open and free is to be exposed and vulnerable. Open space has no trodden paths and signposts. It has no fixed pattern or established human meaning; it is like a blank sheet on which meaning may be imposed. Enclosed and humanized space is place. Compared to space, place is a calm center of established values. Human beings require both space and place. Human lives are a dialectical movement between shelter and venture, attachment and freedom.
> **YI-FU TUAN**[4]

A perspective on place that rests on supposed human universals is bound to fail once we encounter an actual place and actual people. To whom, exactly, is place a "calm center of established values"? Not everyone in Maxwell Street. Not everyone anywhere, I suspect.

According to John Agnew, place consists of three elements; location, locale, and sense of place.[5] Location refers to objective position within an agreed spatial framework, such as longitude and latitude. Location allows us to situate ourselves in relation to other locations that are given distances away in a certain direction. Location is the answer to the question "where?"

> The roadways from curb line to curb line of the following streets: West Maxwell Street from the west line of South Union Avenue to the east line of South Sangamon Street, except the roadway of South Halsted Street; West 14th Street and West 14th Place, from the west line of South Halsted Street to the east line of South Sangamon Street.
> **LEGAL DEFINITION OF MAXWELL STREET MARKET,**
> **MUNICIPAL CODE OF CHICAGO**[6]

> IRA BERKOW But what about the street itself?

> TYNER WHITE The street itself. I will say something you should include in your book. Numerically, the street numbering of Maxwell Street and this area is very important. We know first of all that the highest numbered street in Chicago is 138. And we also notice that Maxwell Street is 1300 South and 800 West. 138. Now listen carefully to the word, "Watergate." And imagine, do you hear a 138. This will not be the only example.

Do you remember that the Berlin Wall was constructed on the 13th of August, 1961?

TYNER WHITE, INTERVIEWED BY IRA BERKOW[7]

Nowhere, unless perhaps in dreams, can the phenomenon of the boundary be experienced in a more originary way than in cities. To know them means to understand those lines that, running alongside railroad crossings and across privately owned lots, within the park and along the riverbank, function as limits; it means to know those confines, together with the enclaves of the various districts. As threshold, the boundary stretches across streets; a new precinct begins like a step into the void—as though one had unexpectedly cleared a low step on a flight of stairs.

WALTER BENJAMIN[8]

Locale

Locale, the second component of place, is the physical and social context in which social relations unfold. Locale refers, in one sense, to the landscape of a place—its physical manifestation as a unique assemblage of buildings, parks, roads, infrastructure, and so on. Locale underlines the idea of place as a setting for particular practices that mark it out from other places. These include the everyday practices of work, education, and reproduction, among others. We often know a place, in some sense, as a locale—a unique combination of things and practices within which life unfolds.

The latter facility is an open-air market where, on a busy day, more than a thousand individual proprietors arrange their merchandise on temporary stands, at the tailgate of their trucks, or simply on the pavement, in the expectation of selling some of them to passing pedestrians.

INSTITUTE OF URBAN LIFE[9]

Sense of Place

Sense of place refers to the subjective side of place—the meanings attached to it either individually or collectively.

Another half block and he was past the shops and into real Maxwell Street. The entire area took its name from the great open-air flea market that once lined both sides of Maxwell Street for blocks, packed with immigrants trying to turn that first buck in the new country. Maxwell itself was little more than an alley now, a narrow dusty street of vacant lots and crumbling sidewalks. But along these curbs and back lots for the length of Maxwell and Liberty, and up and down both sides of Peoria and Thirteenth Street, and all the way to the viaducts and railroad tracks to the south, people were selling things.

MICHAEL RALEIGH[10]

The sidewalk merchants, their customers, and the people who come only to watch the action of the open-air market provide a most curious picture of color, motion and noise. It is precisely this concentration of people and activity which causes this area to have an air of excitement and vitality. There is no

question that the excitement in the open-air market is of a special type not found in any other part of the city—or even the Midwest.
INSTITUTE OF URBAN LIFE[11]

Place is not a scale. It can be thought of as much as a way of thinking or approach to the world as an ontological thing in the world. An old rocking chair by the fireplace can be a favorite place to someone who endows it with meaning or associates it with a set of familiar practices. Placeness is thus an attribute of things across scales.

> The Maxwell Street Market is a place, but it is more than a place; some people would call it a tourist attraction, or an institution, because they have never seen any other open-air market where large numbers of vendors compete with one another for trade, and haggle with their customers over price.
> **INSTITUTE OF URBAN LIFE**[12]

> Place can be as small as the corner of a room or as large as the earth itself: that the earth is our place in the universe is a simply fact of observation to homesick astronauts. . . . It is obvious that most definitions of place are quite arbitrary. Geographers tend to think of place as having the size of a settlement: the plaza within it may be counted a place, but usually not the individual houses, and certainly not that old rocking chair by the fireplace.
> **YI-FU TUAN**[13]

Place is not everything—the one scale that place cannot be is the scale of the universe. Place needs something to be outside of it—some other place to be defined in relation to.

Gathering/ Assembling/ Weaving

Another way of thinking about place is as a gathering of things, meanings, and practices.

Such an idea lies at the heart of a long tradition in the discipline of geography—the tradition of regional geography or chorology. In this iteration, geography was and is seen to involve the study of areal differentiation—the way things are different in different places. For much of the discipline's history this was not expressed as differences in "place" but as differences in "region," but the terms referred to more or less the same thing. Geography was about how things came together uniquely in particular places.

> [Geography] interprets the realities of areal differentiation of the world as they are found, not only in terms of the differences of thing from place to place, but also in terms of the total combination of phenomena in each place, different from those at every other place.
> **RICHARD HARTSHORNE**[14]

While various systematic concepts are based on a sense of equivalence across space, place relies on the idea of a coming together of things that occurs once and once only. To some, this kind of thinking marks out geography and history as special, fundamental disciplines—geography dealing with how things coalesce in place, history with how things are gathered temporally.

This focus on the way in which places gather "phenomena" has found different, and more philosophical, expression in recent work.

> Places gather: this I take to be a second essential trait (i.e., beyond the role of the lived body) revealed by a phenomenological topo-analysis. Minimally, places gather things in their midst—where "things" connote various animate and inanimate entities. Places also gather experiences and histories, even languages and thoughts. Think only of what it means to go back to a place you know, finding it full of memories and expectations, old things and new things, the familiar and the strange, and much more besides. What else is capable of this massively diversified holding action? Certainly not individual human subjects construed as sources of "projection" or "reproduction"—not even these subjects as they draw upon their bodily and perceptual powers. The power belongs to place itself, and it is a power of gathering.
> **EDWARD CASEY**[15]

> Maxwell Street is one of the few places left in the City where I can go on a summer Sunday and mix with Black people and White people and Asians and middle-easterners and rich and poor.
>
> I can watch a pick-up blues band, I can eat a pork chop sandwich with grilled onions not in a food court. I can wander, good, I can wander across an open field.
> **SHARON WOLF, 1993**[16]

This fact of gathering makes place complicated as a concept and as a thing in the world. Places are where things (but also memories, emotions, discourse, etc.) gather. This suggests a horizontal drawing of things from the outside—a relation between an inside that gathers and an outside from which things are gathered. It also suggests a constant, dynamic sense of things on the move as they are gathered in place. Places, in this sense, are open rather than closed. They are semipermeable. The fact that places gather leaves open the questions of why and how particular places gather particular things at particular times.

> Thurs, June 29th, 1939—I spent the evening until 11 pm on Maxwell Street, squatting in a stand watching and talking to the bums of the street. Two middle aged, dark-brown negroes lay sprawled across the booth. One had his head pillowed on a profusion of rags. When cars stopped they, both a little crippled, went out to ask if the owners wanted their cars cleaned. Three drunken bums lay curled on the stand, sleeping. The negro named Jim said to his pal—"all those goddam guys do is drink and sleep." Jack is another bum of the street. He is tall, lean, Irish with attractive face and sharp needled black eyes. His face is back around his lips and cheeks with a heavy growth of stubble. It makes his

> face look lean and savage. He has a dirty cap which slants back and front, over his neck, over his eyes. He is a friendly individual. Buying some corned beef and bread he made sandwiches, passed the food on to the negroes, awakened one of the drunks he knew and made him eat a sandwich. A funny little Mexican tramp in a strange hat sauntered by, grinned with a mouth that was missing several teeth in front. He sat on the booth. He was old, hair mixed evenly with grey and discolored black. One of the negroes had a bag that still contained a sandwich. "Want something to eat?" he asked the Mexican and tossed him the bag. Then the negroes laughed loud as the Mexican chewed the hardened sandwich with uncertain teeth.
>
> **WILLARD MOTLEY**[17]

Place has been approached through the metaphor of weaving. This is another way of describing the gathering qualities of place. Here place is imagined as a textile where threads of different elements of the world combine to produce a unique texture. The particularity of place is produced from the things it weaves together.

> A place's "texture" thus calls direct attention to the paradoxical nature of place. Although we may think of texture as a superficial layer, only "skin deep," its distinctive qualities may be profound. . . . Etymologically, texture is associated with both "textile" and "context." It derives from the Latin *texere*, meaning "to weave," which came to mean the thing woven (textile) and the feel of the weave (texture). But it also refers to a "weave" of an organized arrangement of words or other intangible things (context).
>
> **PAUL ADAMS, STEVEN HOELSCHER, AND KAREN TILL**[18]

Robert Sack has taken this idea of place as a loom the furthest. Over several books he has described the way in which place draws together the woof and weft of different realms into a weave. Sack's approach (like that outlined in assemblage theory, discussed below) is firmly realist. The content of place is a mixture of the social, the natural, and the cultural. Realms of society, nature, and meaning are gathered and woven in places. Mapped onto these realms are further domains of thought such as the empirical (mapped onto the natural), the moral (mapped onto the social), and the aesthetic (mapped onto the cultural).

> The best way to model how place functions as a tool is to think of it along the lines of a loom. As something like a loom, place helps us weave together a wide range of components of reality. The weave itself is the landscape and the projects that the place helps support. What does it draw together? In other publications, I have argued that the major components (or spools of thread) come from three *domains*: the empirical, the moral, and the aesthetic. All of these are part of reality, and place helps us weave these empirical, moral, and aesthetic domains together.
>
> **ROBERT DAVID SACK**[19]

If place constantly weaves the social, the natural, and the cultural together, then we need a theory of place that is not consistently deferred to some other theoretical realm such as the social (as in much of critical spatial thought) or the natural (as in the still-powerful tradition of environmental determinism).

> Indeed, privileging the social in modern geography, and especially in the reductionist sense that "everything is socially constructed," does as much disservice to geographical analysis as a whole as privileging the natural in the days of environmental determinism, or concentrating only on the mental or intellectual in some areas of humanistic geography. While one or the other may be more important for a particular situation at a particular time, none is determinate of the geographical. A clear and comprehensive picture of how we are geographical agents requires that we suspend for a while what may be ideological commitments to privilege one set of forces and make geography a part or a consequence of them—in other words, to reduce the geographic to one of these. Suspending this tendency to reduce geography is the only means of providing enough "space" to examine and develop the role of place and space as forces and thus the idea of geographic causality. Once these concepts have room to develop, there is no need for reduction, for the geographic becomes as important to the other elements of the framework as they are to it.
>
> **ROBERT DAVID SACK**[20]

Assemblage

Assemblage theory, the notion that unique and sufficient wholes can be made from an assemblage of parts, should not come as a surprise to geographers.

In *A New Philosophy of Society*, Manuel DeLanda elegantly outlines his approach to assemblage theory. He does this with reference to spaces and places at a range of scales, from the body to the nation-state. DeLanda is certainly not a geographer and appears to be unaware of the last century of theorizing place in geography and beyond. Nonetheless, his assemblage theory is an sophisticated addition to a long tradition of thinking of place in terms of a syncretic combination of diverse elements that form place through their relations in a particular locale.[21]

> Community networks and institutional organizations are assemblages of bodies, but they also possess a variety of other material components, from food and physical labour, to simply tools and complex machines, to the buildings and neighbourhoods serving as their physical locales.
>
> **MANUEL DELANDA**[22]

An assemblage, DeLanda tells us, is a unique whole "whose properties emerge from the interactions between parts."[23] The ways in which these parts are combined are not necessary or preordained but contingent. Individual parts can be removed and become parts of other assemblages. Places are good examples of assemblages.

FIGURE 53.
Corner of Maxwell and Halsted Streets, Chicago, April 1941. Photo by Russell Lee. Library of Congress. LC-USF33-012984-M5. Library of Congress, Prints and Photographs Division, FSA/OWI Collection.

What would replace, for example, "the market" in an assemblage approach? Markets should be viewed, first of all, as concrete organizations (that is, concrete market-places or bazaars) and this fact makes them assemblages made out of people and the material and expressive goods people exchange.
MANUEL DELANDA[24]

It's a throwback to the Tower of Babel with the haggling over process being carried on in a dozen tongues. Barkers, pitchmen and spielers shout their wares over the blare of radios perched on windowsills of dirty, tired-looking buildings elbowing each other like the stands on the street.

The screech of live chickens adds to the din of organized confusion....

From a side street comes the high-pitched voice of a revivalist, wailing a hymn into a microphone. A man sits cross-legged on the sidewalk—adjusting the dials for the loud-speaker while swaying in rhythm to the song.

Maxwell is a street of a thousand smells. The pungent odor of garlic, sizzling hot-dogs, spoiling fruit, aging cheese and pickled fish blend in a unique aroma.
***CHICAGO SUN-TIMES*, 1951**[25]

A familiar example. My home is a place. It is made from red bricks, breeze blocks, terra-cotta tiles, windowpanes, copper wires, plastic outlets, wooden floorboards, cotton curtains, a stainless-steel hob and oven, mortar and glue, all the things we eat, notes on the fridge. The list could go on for a long time, its contents together constituting my home. The way the disparate items are assembled make my home different from a supermarket or a football stadium and even, in details at least, neighboring homes. The place as assemblage is not just a random collection of parts, nor is it static. My home is always changing: the food in the fridge is rarely the same as the day before, cracks grow in the plaster, weeds push up between paving stones in the back garden. I have considered removing the current floorboards and putting in new ones. The assemblage remains despite the changes in parts. My home is a discrete thing—an assemblage—comprising relations between parts and the kinds of things we do with those parts. All places can be thought of in this way.

While many elements can be removed and replaced in the assemblage that is "my home," one element that cannot be removed and replaced is me. In fact, the home referred to above is no longer my home. That material structure still exists in London—no doubt adjusted to the tastes of new owners. I now live in West Hartford, in Connecticut, in the United States. While place is certainly an assemblage, it needs an "I" or a "we" at the center of it. For each I or we there is something distinct—something that makes the whole greater than the sum of the parts.

> While the decomposition of an assemblage into its different parts, and the assignment of a material or expressive role to each component exemplifies the analytic side of the approach, the concept of territorialization plays a synthetic role, since it is in part through the more or less permanent articulations produced by this process that a whole emerges from its parts and maintains its identity once it has emerged.
> **MANUEL DELANDA**[26]

Central to DeLanda's notion of assemblage are two dimensions, or axes, which he derives from Deleuze and Guattari. One axis concerns the role played by components of a whole. It is an axis that has *expressive* roles at one end and *material* roles at the other. The second axis concerns the degree to which assemblages are stabilized around the coherent identity or, alternatively, are destabilized and made unclear.

> The former are referred to as processes of *territorialization* and the latter as processes of *deterritorialization*. One and the same assemblage can have components working to stabilize its identity as well as components forcing it to change or even transforming it into a different assemblage. In fact, one and the same component may participate in both processes by exercising different sets of capacities.
> **MANUEL DELANDA**[27]

Territorialization/Deterritorialization

If we return to the assemblage and place that is my house, we can see that there are forces at work that stabilize its identity. These range from the laws that make it my property to the maintenance that prevents its falling down. At the same time, there is entropy.

> Take, for example, the building you walk through/within—what is the speed of flux that is keeping it assembled? It seems permanent, less ephemeral than you, but it is ephemeral nonetheless: whilst you are there it is falling down, it is just happening very slowly (hopefully).
> **J. D. DEWSBURY**[28]

> The concept of territorialization must be first of all understood literally. Face-to-face conversations always occur in a particular place (a street-corner, a pub, a church), and once the participants have ratified one another a conversation acquires well-defined spatial boundaries. Similarly, many interpersonal networks define communities inhabiting spatial territories, whether ethnic neighbourhoods or small towns, with well-defined borders. . . . So in the first place, processes of territorialization are processes that define or sharpen the spatial boundaries of actual territories. Territorialization, on the other hand, also refers to non-spatial processes which increase the internal homogeneity of a neighbourhood. Any process which either destabilizes spatial boundaries or increases internal heterogeneity is considered deterritorializing.

> The notion of the structure of a space of possibilities is crucial in assemblage theory given that, unlike properties, the capacities of an assemblage are not given, that is, they are merely possible when not exercised. But the set of possible capacities of an assemblage is not amorphous, however open-ended it may be, since different assemblages exhibit different sets of capacities.
> **MANUEL DELANDA**[29]

Places both gather and disperse. They collect things from outside and are thus constituted through their relations to the world beyond. But things are always also escaping place. Places, therefore, are in process. They are becoming and dissolving on a daily basis.

A place, by definition, is differentiated from the other places that are elsewhere. It is also connected to them. Places are "elsewhere" in relation to other places. Two or more places may exist in the same location at the same time. The way I experience downtown Hartford, for instance, is certainly different from the way a recent immigrant from Puerto Rico experiences downtown Hartford.

Thinking of the world in terms of places means distinguishing one thing from another. It is way of naming heterogeneity and delineating spatial difference. The world is made up of heterogeneous places. And yet, places are not internally homogeneous. A place is made up of gathered things, meanings, and practices that are different from one another. The unique assemblage of differences makes the place what it is and differentiates it from other unique assemblage-places.

> We can, of course, grasp places (even from within the very place so grasped) as having a character and identity of their own. And this is so not only in

virtue of the way a particular place allows things to appear within it, but also in terms of the way in which any such place is always itself positioned in relation other places and provides a certain "view" of such places. Places are thus internally differentiated and interconnected in terms of the elements that appear within them, while they also interconnect with other places—thus places are juxtaposed and intersect with one another; places also contain places so that one can move inwards to find other places nested within a place as well as move outwards to a more encompassing locale.

JEFF MALPAS[30]

There is something paradoxical about this state of affairs—this combination of internal and external heterogeneity. How do we delineate a particular set of differences and call it a particular place without such a definition being arbitrary? How is it that Maxwell Street can be Maxwell Street beyond the simple act of naming it?

Materialities/ Meanings/ Practices

What is it that is gathered in place? We have listed as necessary ingredients for place materialities, meanings, and practices. Each interacts with the others, and each is necessary for a satisfactory account of place.

> Despite the character resulting from the unique series of façades along Maxwell Street and the color and life emanating from the juxtaposition of a multitude of signs, sidewalk tables, hawkers and an infinite variety of merchandise, the preponderant number of buildings and temporary stalls along the street area are dilapidated and have long since outlived their usefulness. Thus, it appears that a new physical setting for the Maxwell Street Market should be created and preferably in conjunction with the proposed relocation of the shopping center.
>
> **CHICAGO DEPARTMENT OF URBAN RENEWAL**[31]

Materialities

Places have a material presence. What we think of when we think of a place we have not been are perhaps some of the things that constitute it, particularly its buildings and monuments. New York has the Empire State Building, Paris the Eiffel Tower. Every place has a material (solid, concrete) landscape, some remarkable and some less so.

> The largest impression is one of a junked-up neighborhood—the stands—basements and lofts filled with nondescript things. . . .
>
> The architecture and structure of the brick building show that this was once a first class neighborhood. . . .
>
> The buildings kneel to the street.
>
> **WILLARD MOTLEY**[32]

There are many things in Place Saint-Sulpice; for instance: a district council building, a financial building, a police station, three cafés, one of which sells tobacco and stamps, a movie theatre, a church on which La Vau, Gittard,

FIGURE 54.
Shop on Maxwell Street, Chicago, April 1941. Photo by Russell Lee. Library of Congress. LC-USF33-012984-M2. Library of Congress, Prints and Photographs Division, FSA/OWI Collection.

> Oppernord, Servandoni, and Chalgrin have all worked, and which is dedicated to a chaplain of Clotaire II, who was bishop of Bourges from 624 to 644 and whom we celebrate on 17 January, a publisher, a funeral parlor, a travel agency, a bus stop, a tailor, a hotel, a fountain decorated with the statues of four great Christian orators (Bossuet, Fénelon, Fléchier, and Massillon), a newsstand, a seller of pious objects, a parking lot, a beauty parlor, and many other things as well.
> **GEORGES PEREC**[33]

But places have more of a material presence than just their landscape. They also have the multitude of things that pass through them. My home as a place has all the stuff of my family's everyday life—books, toys, food, waste, souvenirs, flowers bought for the weekend and then added to the compost. Urban places, and particularly marketplaces, connected as they are to the wide world beyond, are teeming with things that do not stay long.

> Wire-fencing in various size rolls; lighted compasses; a bench drill press; spices in industrial-sized containers; refrigerator/freezers; glazed ceramic tiles in boxes; a telephone-answering machine; long-stemmed glasses and crystal goblets; gloves (ski and regular); automobile wheels and tires; underwear; jeans; jackets and jump suits; battery chargers; a snow blower; notebooks and paper for school; baseball trophies; skates (ice and roller); and comforters.
> ***CHICAGO TRIBUNE*, 1981**[34]

Things suggest tangibility. They tend toward the solid. Things differ from objects in that they have not yet been represented or "objectified."

> There are many intersections in the ways of ongoing flux, places of steady but impermanent homeostasis.
>
> These are called things.
> Thing (Old English): an assembly, a gathering
> Thingan (Old English): to invite, to address
> Althing (Icelandic): the parliament
>
> **DON MCKAY**[35]

Things travel. They circulate within and beyond places. As they do so they acquire and discard meaning. Things in a place help make that place what it is, in part because of what they bring from elsewhere.

> In doing the biography of a thing, one would ask questions similar to those one asks about people: What, sociologically, are the biographical possibilities inherent in its "status" and in the period and culture, and how are these possibilities realized? Where does the thing come from and who made it? What has been its career so far, and what do people consider to be an ideal career for such things?
>
> **IGOR KOPYTOFF**[36]

In addition to the materialities of the built landscape, and the multitude of objects that exist in and pass through places, there are other kinds of materialities at work in place. It is important not to conflate the material with the solid, tangible, concrete. Invisible currents of air, for instance, are every bit as material as concrete and also contribute to the gathering of place.

> Thus it seems that we have human minds on the one hand, and a material world of landscape and artefacts on the other. That, you might think, should cover just about everything. But does it? Consider, for a moment, what is left out. Starting with landscape, does this include the sky? . . . How about sunlight? Life depends on it. But if sunlight were a constituent of the material world, then we would have to admit not only that the diurnal landscape differs materially from the nocturnal one, but also that the shadow of a landscape feature, such as a rock or tree, is as much a part of the material world as the feature itself. For creatures that live in the shade, it does indeed make a difference. What, then, of the air? When you breathe, or feel the wind on your face, are you engaging with the material world? When the fog descends, and everything around you looks dim and mysterious, has the material world changed, or are you just seeing the same world differently? Does rain belong to the material world, or only the puddles that it leaves in ditches and potholes? Does falling snow join the material world only once it settles on the ground? As engineers and builders know all too well, rain and frost can break

up roads and buildings. How then can we claim that roads and buildings are part of the material world, if rain and frost are not?
TIM INGOLD[37]

Places are made up of materials and their properties. These materials, which are gathered, include more than landscape and artifacts. Smells, for instance, form characteristic parts of a place, as invisible stimulants interact with the sensors in our nose and brain. Waves of sounds, like smells, are part of the assemblage that gathers to form distinctive places.

I'm fascinated by relationships of bells and music, for example, a church bell with the same resonant decay time as one of the oldest organs in Finland. Or bells and space, for example, when walking with a shepherd in Italy and hearing, a kilometer away, the funeral bells from the church overlapping the bells of his 50 sheep. These are historically layered relationships in sound, like the way belled flocks move through the countryside, making place audible. . . . Or the way time bells and chimes make communities audible. And the particularity of interactions of these kinds of bell sounds with cars and motorbikes, with televisions and radios, and all the sounds of the modern world.
STEVEN FELD AND DONALD BRENNEIS[38]

While landscape is shaped through senses in combination, the particular qualities of sound, especially of the being-heard rather than being-seen, alert us to landscape's shaping of and through forms of address, mark landscape as colloquial.
DAVID MATLESS[39]

Barkers, spielers, pitchmen, and hucksters shout their wares while radios boom and customers haggle in a dozen languages. Merchandise drapes from awnings, spills over sidewalk stands and creaking pushcarts, litters the pavement and walks wherever the hawkers elect to take their stand. There is the sharp odor of garlic, sizzling redhots, spoiling fruit, aging cheese, and strong suspect smell of pickled fish. Everything blends like the dazzling excitement of a merry-go-round.
***CHICAGO SUNDAY TRIBUNE*, 1947**[40]

Put another way, we cannot simply rein things in and root them. It is not enough to use the "material" and "materiality" in such a way as to invoke a realm of reassuringly tangible or graspable objects defined against a category of events and processes that apparently lack "concreteness." Rather, we only begin to properly grasp the complex realities of apparently stable objects by taking seriously the fact that these realities are always held together and animated by processes excessive of form and position.
ALAN LATHAM AND DEREK MCCORMACK[41]

One way of approaching place is to describe its visible material contents—to provide a kind of catalog of place. There would be some theoretical and poetic merit in such an attempt, if only in its revealing the foolhardiness of the project. At some point interpretation becomes part of the endeavor. Places are also places because of the meanings produced through acts of representation. Places are both represented (by poets, photographers, politicians, etc.) and are themselves representational.

> The contents of culture can be itemized, although if one is meticulous, the list threatens to grow to interminable length. Culture is not such a list. Landscape, likewise, is not to be defined by itemizing its parts. The parts are subsidiary clues to an integrated image. Landscape is such an image, a construct of the mind and of feeling.
>
> **YI-FU TUAN**[42]

> In the arcades, bolder colours are possible. There are red and green combs.
>
> Preserved in the arcades are types of collar studs for which we no longer know the corresponding collars or shirts.
>
> Should a shoemaker's shop be neighbour to a confectioner's his display of shoelaces will start to resemble licorice.
>
> One could imagine an ideal shop in an ideal arcade—a shop which brings together all métiers, which is doll clinic and orthopedic institute in one, which sells trumpets and shells, birdseed in fixative pans from a photographer's darkroom, ocarinas as umbrella handles.
>
> **WALTER BENJAMIN**[43]

Meaning

Materialities are attached to meanings and vice versa. Meaning has always been central to the geographical analysis of place. Places are locations with meaning. Things, smells, sounds—all become part of worlds of meaning as humans attempt to make sense of them. The ways in which this happens are almost infinite.

Nominalism—the act of naming—is an important part of the creation of meaning in place. Simply naming a place can cause it to be a place rather than a mere location.

> Generic terms are not as powerful evocators of place as are proper names. To call a feature in the landscape a "mount" is already to impart to it a certain character, but to call it "Mount Misery" is to significantly enhance its distinctiveness, making it stand out from other rises less imaginatively called. The proper name and the geographical feature so merge in the consciousness of the people who know both that to change the name is to change, however subtly and inexplicably, the feature itself.
>
> **YI-FU TUAN**[44]

> The true expressive character of street names can be recognized as soon as they are set beside reformist proposals for their normalization.
>
> **WALTER BENJAMIN**[45]

> You might look for a whole week by consulting lampposts on the West Side and you wouldn't find "Maxwell Street." The hand of the iconoclast has been busy with West Side tradition and the street to which the famous police station gave dignity is simply and flatly West Thirteenth Place.
>
> But the Maxwell Street Station is still there, and the street is still called Maxwell Street by everybody but the postman.
>
> ***CHICAGO TRIBUNE*, 1895**[46]

Places are sites where stories gather. Individual stories (telling a story without a place would be experimental in the extreme), collective stories, official stories, subversive stories.

> Narratives about people's places in places continuously materialize the entity we call place. In its materializations, however, there are conflicts, silences, exclusions. Tales are told and their meanings wobble and shift over time. Multiple claims are made. Some stories are deemed heretical. The resulting dislocations, discontinuities, and disjunctures work to continually destabilize that which appears to be stable: a unitary, univocal place.
>
> **PATRICIA PRICE**[47]

Perhaps parataxis is one way to destabilize narrative—a way particularly appropriate to the assemblage of place.

Practice

In addition to materials and meanings, places are marked by practice. People, in combination with various objects, are always doing things. To many observers this has been the most important defining feature of place—a kind of choreography of deeply engrained habits and semi-organized rhythms that makes a place distinctive. Places gather practices. They are lived.

> The widespread view of the city as a built form rather than as a lived form stems from a strange logical fallacy with a fittingly strange name. *Reification* comes from *red*, Latin for "thing," which has led some to suggest instead the silly-sounding thingification.
>
> **DAVID KISHIK**[48]

In Georges Perec's book *An Attempt at Exhausting a Place in Paris*, he sits at a number of a café tables at the Place Saint-Sulpice, noting the banal happenings outside. He notes the "things" that make up the square and how they have been previously inventoried and accounted for. It is not the buildings he is aiming to capture but the rest of what makes up place: "that which is not noticed, that which has no importance; what happens when nothing happens other than the weather, people, cars and clouds."[49]

There is a long history of writers engaging with place as a *choreography* of activity. Some notable examples are Jane Jacobs's account of sidewalk life in Manhattan, Michel de Certeau's urban stories, David Seamon's description

FIGURE 55.
Maxwell Street Market sign, now on Desplaines Street. Photo by author.

of "place-ballets," and Henri Lefebvre's theorization of rhythm while looking down on a Paris square.[50]

> Maxwell:—its 12:30 at night. Two youths have driven in from southern Illinois to buy white mice for a circus game. Places closed. Ask me when they open. Then ask cop. Crippled negro staggers down middle of street. An auto almost hits him. He yells—"kiss my ass!" at the top of his voice. Another car has to swerve to avoid him. Again he yells. Cop—not 10 feet away, keeps talking to the youth as if he hasn't seen or heard the negro. Anything goes.
>
> Sight-seeing buses pass slowly down Maxwell at night. The tourists gawk. There is always a shouted chorus of "Rubberneck" to greet them. Sometimes the youths of the neighborhood collect rotten tomatoes and other vegetables and fruit and redden the windows of the buses.
>
> Maxwell Street—the prostitute is treated like a human being down there—as just another merchant. Everyone speaks to her, laughs, talks, jokes with her. The men merchants aren't afraid of being seen standing on the corner talking to her in the daylight. . . .
>
> The stout black woman sometimes dressed in a nun's outfit, sometimes all in white, walks down the middle of Maxwell dressed in a loud purple graduate's gown and a black mortar cap. She generally carries a can for contributions to her church.
>
> At 1am each morning (2 am Sunday mornings) when the taverns close the prostitutes come out on the streets again to hustle customers shoved from the taverns.
>
> The market extends between low, dilapidated tenements, the 1st floors of which are used as stores and shops of all descriptions. What space is left in the middle of the street is thronged with unending streams of pedestrians—a surging mass of bargain hunters. Weary old men, haggard women stand behind the carts.
>
> **WILLARD MOTLEY**[51]

> The stretch of Hudson Street where I live is each day the scene of an intricate sidewalk ballet. I make my own first entrance into it a little after eight when I put out the garbage can, surely a prosaic occupation, but I enjoy my part, my little clang as the droves of junior high school students walk by the centre of the stage dropping candy wrappers. . . . While I sweep up the wrappers I watch the other rituals of the morning: Mr. Halpart unhooking the laundry's handcart from its mooring to a cellar door, Joe Cornacchia's son-in-law stacking out the empty crates from the delicatessen, the barber bringing out his sidewalk folding chair, Mr. Goldstein arranging the coils of wire which proclaim the hardware store is open, the wife of the tenement's superintendent depositing her chunky three-year-old with a toy mandolin on the stoop, the vantage point from which he is learning the English his mother cannot speak. Now the primary children, heading for St. Luke's, dribble through to the west, and the children from P.S. 41 hearing towards the east.
>
> **JANE JACOBS**[52]

The ways in which a place emerges from practice are never entirely predictable but do have some measure of order. Some practices recur on a regular timetable (going to work or school) while others are less predictable (the skateboarder, the protester). Some kinds of practice appear to conform to what Allan Pred has called "dominant institutional projects," while others seem to oppose them. Still others simply meander around in irreducible ways.[53]

> "To dwell" as a transitive verb—as in the notion of "indwelt spaces"; herewith an indication of the frenetic topicality concealed in habitual behavior. It has to do with fashioning a shell for ourselves.
>
> **WALTER BENJAMIN**[54]

> On 14th Street a small card table, manned by two men in their mid-twenties, was filled with magic tricks and novelties. These men stated that although they had been operating for only six weeks, they have attended every Saturday and Sunday during this time. Both of these men have other jobs which are their principal sources of income. One of the men is also a magician and on occasion performs for various groups for pay. Certainly the brisk pace of sales at this stand was primarily due to the fact that one of the young men was performing magic tricks to induce customers to stop.
>
> **INSTITUTE OF URBAN LIFE**[55]

Practices cannot be separated easily from materialities. A practice repeated in a different place can hardly be called the same practice. Walking down Maxwell Street today is hardly the same thing as walking down Maxwell Street in 1882 or 1953. None of those are the same as walking through Washington Square in New York City or down an alleyway in a market in Morocco.

Places gather materialities, meanings, and practices. Together, these produce unique assemblages. Territorializing forces pull them together into tight, bounded knots that produce areal differentiation and the possibility of difference. Deterritorializing forces work to pull them apart. This process of territorialization and deterritorialization works across a horizontal plain of flows and connections.

Roots/Routes

> Universe and place are connected in a knot as difficult to form as to imagine. On the one hand, the local sees obstructions on its borders, causing neighbouring areas to be inaccessible; the extremal path on the other hand knows no obstacle and recognizes no place.
>
> **MICHEL SERRES**[56]

The emphasis on place as a singular collection of things in the natural and human worlds has traveled hand in hand with a focus on the vertical axis of rootedness and belonging. Such approaches to place are indebted to Heidegger and his insistence on the processes through which people become rooted. This sense of rootedness is achieved through being in place, building, and dwelling. Heidegger's famous log cabin in the Black Forest links him to the cosmos and to the earth as nature. The important axis here is a vertical one.

> Let us think for a while of a farmhouse in the Black Forest, which was built some two hundred years ago by the dwelling of peasants. Here the self-sufficiency of the power to let earth and heaven, divinities and mortals enter in simple oneness into things, ordered the house. It placed the farm on the wind-sheltered mountain slope looking south, among the meadows close to the spring. It gave it the wide overhanging shingle roof whose proper slope bears up under the burden of snow, and which, reaching deep down, shields the chambers against the storms of the long winter nights. It did not forget the altar corner behind the community table; it made room in its chamber for the hallowed places of childbed and the "tree of the dead"—for that is what they call a coffin there: the Totenbaum—and in this way it designed for the different generations under one roof the character of their journey through time. A craft which, itself sprung from dwelling, still uses its tools and frames as things, built the farmhouse.
> **MARTIN HEIDEGGER**[57]

Central to most approaches to place is the sense that it is unique and particular—"here" is separate and distinguished from "there." Location, locale, and sense of place all contribute to this sense of singularity, as does the unique gathering of material, meaning, and practice. This is one aspect of place—its vertical axis—a sense of depth and boundedness.

But places also exist on a horizontal axis. They are connected to other places and, in part, derive their identities from these connections. They overspill their bounds, giving them width and a distributed presence.

> On a busy Sunday, as many as a thousand of these stands may be in operation, a number which results in the spreading of the market to adjoining streets not covered by the legal definition, and utilizing almost double the area provided by the municipal ordinance.
> **INSTITUTE OF URBAN LIFE**[58]

> Home was the centre of the world because it was the place where a vertical line crossed with a horizontal one. The vertical line was a path leading upwards to the sky and downwards to the underworld, the horizontal line represented the traffic of the world, all the possible roads leading across the earth to other places. Thus, at home, one was nearest to the gods in the sky and the dead in the underworld. This nearness promised access to both. And at the same time, one was at the starting point and, hopefully, the returning point of all terrestrial journeys.
> **JOHN BERGER**[59]

> Maybe the first [lesson] was that the very term place is problematic, implying a discrete entity, something you could put a fence around. . . . What we mean

by place is a crossroads, a particular point of intersection of forces coming from many directions and distances.
REBECCA SOLNIT[60]

Place is latitudinal and longitudinal within the map of a person's life. It is temporal and spatial, personal and political. A layered location replete with human histories and memories, place has width as well as depth. It is about connections, what surrounds it, what formed it, what happened there, what will happen there.
LUCY LIPPARD[61]

Places are (in part) relational. They are produced through relations with elsewhere. We have both an idea of dwelling based on the habitual repetition of "here" and the idea of a place that articulates processes that are always beyond it.

The difficulty in reflecting on dwelling: on the one hand, there is something age-old—perhaps eternal—to be recognized here, the image of the abode of the human being in the maternal womb; on the other hand, this motif of primal history notwithstanding, we must understand dwelling in its most extreme form as a condition of nineteenth century existence.
WALTER BENJAMIN[62]

Instead then, of thinking of places as areas with boundaries around, they can be imagined as articulated moments in networks of social relations and understandings, but where a large proportion of those relations, experiences and understandings are constructed on a far larger scale than what we happen to define for that moment as the place itself, whether than be the street, or a region or even a continent. And this in turn allows a sense of place which is extroverted, which includes a consciousness of its links with the wider world, which integrated in a positive way the global and the local.
DOREEN MASSEY[63]

Around Halsted and Maxwell Streets swarmed the Chicago ghetto, with many thousands of Jews from Russia and eastern Europe, who spoke no English, who arrived with tickets pinned to their clothes and placards hung around their necks. Street and tenements teemed with bearded patriarchs and their families. Stores sprang up with pullers-in on the walks and everywhere were pushcarts which sold everything from garlands to garlic to women's drawers. The Jewish wave followed closely on the Irish, and with Blue Island Avenue the dividing line, there were wonderful ruckuses that kept the Maxwell Street Station house humming.
JACK LAIT AND LEE MORTIMER[64]

Maybe instead of, or as well as, the time-embeddedness that enables that relational achievement of the establishment of a (provisional) ground, such histories push a need to rethink our security. Certainly such histories have the potential to be read as removing the absoluteness of such grounding, so that all we are left with is our interdependence, a kind of suspended, constantly-being-made interdependence, human and beyond human.

DOREEN MASSEY[65]

This emphasis on the horizontal plain of connections in place raises some questions that have not been satisfactorily grappled with. There is little in the formulation of a progressive sense of place to tell us why places become connected in the way they do. Is there something "there"—in place—that leads to these connections? Are there qualities of place that make it more or less likely to connect in particular ways? One answer to this is to point to previous networks and connections that are likely to have created a context for present ones. This, however, simply delays answering the question, perhaps indefinitely. The question (and the answer) raise the question of temporality.

Temporalities

It is not that what is past casts its light on what is present, or what is present its light on what is past; rather, image is that wherein what has been comes together in a flash with the now to form a constellation.

WALTER BENJAMIN[66]

All the stories of Progress, of Development, of Modernization (such as the movement from traditional to modern, or the Marxist progression through modes of production (feudalism, capitalism, socialism, communism) and of many formulations of the story of "globalization") . . . share a geographical imagination that involves this manoeuver: it rearranges spatial differences into temporal sequence. Such a move has enormous implications: it implies that places are not genuinely different . . . but simply "behind" or "advanced" within the same story; their "difference" consists only of their place in the queue.

DOREEN MASSEY[67]

The Maxwell Street Market, existing to great extent without buildings, is a temporal environment more than a geographical one that has gained some permanence as an Old World caravan which is not limited by its physical setting or buildings.

CHICAGO DEPARTMENT OF PLANNING, 1989[68]

Places are always changing. And yet, in even the most rapidly changing place, there is a sense of continuity. In addition to the vertical axis of here-ness and the horizontal axis of flows and connections, we also need to think about time. How does a place at a moment in time (say 1920) relate to what is assembled at the same objective location years later (say 2019)? A theory of place needs an account of time and change.

"Ain't seen Sam. Not in a while."

"But you are a friend of his."

"Yeah. Don't see him much now."

"Does he still set up on Maxwell Street?"

Brown stared at him. "Far as I know. I don't go down there no more myself. I'm too old to stand out there all day for a couple dollars."

"You used to sell things down there?"

"Yes, sir. Just one more way to get by. Tryin' to make a dollar. Sam, he didn't have much to sell, last I heard." O.C. looked out the little window. "Some folk down there just be selling junk, just trying to get by. It's not what it used to be, Maxwell Street."

"I remember."

"Gonna be gone soon, all of it."

MICHAEL RALEIGH[69]

This third axis of place, temporality, is itself an effect of verticality and horizontality. Traditional accounts of dwelling in a singular place locate the place's identity in its (singular) history—"this is where x happened." Dwelling is dwelling in time as much as in place. It is often this notion of temporality that forms the basis of museums, memorials, and heritage parks. But temporalities are also effects of the way places act as nodes for mobilities and other kinds of flow. Duration can be a relational achievement.

In spite of its prosperity, the Rialto of the ghetto—Maxwell Street—is fast passing away. As the immigrants get into closer touch with the outside world, they see that after all the ghetto offers but limited opportunities for success. They establish themselves in stores and offices in other parts of the city and become large-scale merchants, real estate dealers, manufacturers, and building contractors. Compared again with the world beyond the ghetto, the ghetto world shrinks to a vanishing point. Not only do the Jewish merchants move away from Maxwell Street to more reputable quarters, but in recent years there have been few recruits to fill the vacancies. A few recent immigrants still drift to the push-carts, but generally only for a short time, until they have accumulated sufficient wealth to move elsewhere. Maxwell Street is declining, and is being left to the rats that haunt its streets at night.

LOUIS WIRTH[70]

Materiality/ Temporality

Places are assemblages, gatherings, of materiality, meaning, and practice. Each of these have temporal dimensions. At first glance (at least to me) the material world has the best chance of enduring. Solid buildings, roads, and parks, once in place, gather inertia. Only the most powerful of forces can erase them. Meanings, insubstantial things, can easily change, and yet we know that some collective meanings become remarkably intransigent—despite the attempts at place-branding. It is not easy to transform a place with negative meanings into a place with positive ones. Practices appear to be the most fleeting of all. Each occurrence of a practice, each thing that happens in a place, happens in exactly that way only once. Its singularity is impossible to capture. And yet, some practices become habits and routines. When repeated over and over (in approximately the same way) they create as powerful a sense of place as any.

> Habits throw themselves together into an aesthetics. Townies leave windows un-curtained and open through the night and in the deadest cold of the winter.
>
> People walk the neighborhood to peer into the scenes of people reading the paper at night or up early drinking their coffee. Lamps are favored over overhead lighting, lending texture and specificity to ordinary, no-big-deal living.
>
> **KATHLEEN STEWART**[71]

Solid landscapes appear to be enduring. Limestone, slate roofs, red bricks, mortar—however weathered and suffering from entropy—convey a sense of being there for the long term.

> Buildings stabilize social life. They give structure to social institutions, durability to social networks, persistence to behavior patterns. What we build solidifies society against time and its incessant forces for change. . . . Brick and mortar resist intervention and permutation, as they accomplish a measure of stasis. And yet, buildings stabilize imperfectly. Some fall into ruin, others are destroyed naturally or by human hand, and most are unendingly renovated into something they were not originally.
>
> **THOMAS F. GIERYN**[72]

> 719 W. Maxwell Street
>
> This building was originally constructed by Mrs. Roma Stein as a two story brick store and dwelling in 1881. In 1903, then owners Farber and Wittenberg, who also owned the building next door, hired Alexander L. Levy to design a one-story addition for the building Since that time the building had numerous other alterations including the removal of the third story in 1934. Although the building retains some architectural details, it no longer conveys its historic appearance and would likely be deemed a non-contributing feature to a potential historic district.
>
> **ARCHAEOLOGICAL SURVEY OF THE MAXWELL STREET AREA, 1994**[73]

It is exactly this stubborn nature of the solid concreteness of the built landscape that makes it a potential nuisance for the smooth circulation of capital. One way of thinking about this solid aspect of place is as "fixed capital" or even as "dead labor." Architects and others refer to "embodied energy"—the sum total of all the energy that went into making a building.

> The produced geographical landscape constituted by fixed and immobile capital is both the crowning glory of past capitalist development and a prison that inhibits the further progress of accumulation precisely because it creates spatial barriers where there were none before. The very production of this landscape, so vital to accumulation, is in the end antithetical to the tearing down of spatial barriers and the annihilation of space by time.
>
> **DAVID HARVEY**[74]

It is not just Marxists who ascribe to the "material" a power of intransigence. Outside of a Marxist framework, the durability of "things" has been credited with the ability to ensure "duration."

> Everything in the definition of macro social order is due to the enrolment of nonhumans . . . even the simple effect of duration, of long-lasting social force, cannot be obtained without the durability of nonhumans to which human local interactions have been shifted.
>
> **BRUNO LATOUR**[75]

The past continues to exert its influence on the present through this material presence and the inertia it represents. And yet even the most solid part of place is, in fact, (slowly) melting into air. Even a piece of the landscape as apparently unchanging as a public monument was (and is) just a very protracted "event."

> At first sight we should hardly call this an event. It seems to lack the element of time or transitoriness. But does it? . . . The static timeless element in the relation of Cleopatra's Needle to the Embankment is a pure illusion generated by the fact that for purposes of daily intercourse its emphasis is needless. What it comes to is this: Amidst the structure of events which form the medium within which the daily life of Londoners is passed we know how to identify a certain stream of events which maintain permanence of character, namely the character of being the situations of Cleopatra's Needle. Day by day and hour by hour we can find a certain chunk in the transitory life of nature and of that chunk we say, "There is Cleopatra's Needle." If we define the Needle in a sufficiently abstract manner we can say that it never changes. But a physicist who looks on that part of the life of nature as a dance of electrons, will tell you that daily it has lost some molecules and gained others, and even the plain man can see that it gets dirtier and is occasionally washed. Thus the question of change in the Needle is a mere matter of definition. The more abstract your definition, the more permanent the Needle. But whether your Needle change or be permanent, all you mean by stating that it is situated on the Charing Cross Embankment, is that amid the structure of events you know of a certain continuous limited stream of events, such that any chunk of that stream, during any hour, or any day, or any second, has the character of being the situation of Cleopatra's Needle.
>
> **ALFRED NORTH WHITEHEAD**[76]

> Are they saying that there has always been a stand at this particular location since 1890 or 1940? Or are they saying that the actual physical building standing today is what was built in 1890 or 1940? I think this would be difficult to prove. These stands are not built as permanent buildings—I'm sure they are completely rebuilt from top to bottom over the space of 15 to 20 years. If you cannot verify the date of the actual materials, then I would recommend not including them in the resource count.
>
> **ANN SWALLOW, 1994**[77]

Buildings, and other elements of the material landscape of place, are also linked to other places along the horizontal plane of connections. While buildings may appear to derive their sense of relative permanence from their singularity they are in fact achievements of networks as much as they are singular edifices. They are connected to the world through literal material connections such as sewers and telephone lines. The rock they are made of comes from elsewhere. Even the aesthetics of the façade are likely to have traveled.

> The mutability of Skin seems to be accelerating. Demographer Joel Garreau says that in "edge cities" (new office and commercial developments on the periphery of older cities) developers are accustomed to fine-tune their buildings by changing rugs and façades—a typical "façadectomy" might go upscale from pretentious marble veneer to dignified granite veneer to attract a richer tenant. Developers expect their building Skins to "ugly out" every fifteen years or so, and plan accordingly.
>
> **STEWART BRAND**[78]

> The building becomes a place where a number of material and immaterial links meet in a node of relations, whose durability is both relative and negotiated. It is in this way that the building is able to engage and negotiate with a number of disparate realms. . . . The building as a permeable entity becomes less an individual building block in a collection of blocks . . . [than] an unstable assemblage that is intimately connected to and renegotiated by the surrounding buildings, streets, communities, and economies and the world beyond.
>
> **LLOYD JENKINS**[79]

This connectedness across a horizontal plain can also be thought of, following Whitehead, as an "event" that is always subject to DeLanda's territorializing and deterritorializing pressures.

> The materiality of the building is a relational effect, its "thing-ness" is an achievement of a diverse network of associates and associations. It is what we might think of as a building event rather than simply a building. Conceived of in this way, a building is always being "made" or "unmade," always doing the work of holding together or pulling apart.
>
> **JANE M. JACOBS**[80]

Meaning/Temporality

Places do not endure just through their material presence. The world of meaning is also important to the temporality of place. Sometimes the stories we tell about a place continue in more or less the same form long after the material landscape has changed. Place is powerfully linked to memory through the connections between materiality and meaning.

> It is the stabilizing persistence of place as a container of experiences that contributes so powerfully to its intrinsic memorability. An alert and alive memory connects spontaneously with place, finding in it features that favor and

> parallel its own activities. We might even say that memory is naturally place oriented or at least place-supported.
>
> **EDWARD CASEY**[81]

As with the materiality of place, the persistence of memory is not purely a product of "here" but is produced through interactions between the inside and the outside that constitute it. It has both vertical and horizontal aspects.

> Place memory encapsulates the human ability to connect with both the built and natural environments that are entwined in the cultural landscape. It is the key to the power of historic places to help citizens define their public pasts: places trigger memories for insiders, who have shared a common past, and at the same time places often can represent shared pasts to outsiders who might be interested in knowing about them in the present.
>
> **DELORES HAYDEN**[82]

Meaning, and the narratives, stories, ideologies, and memories that produce it, does not simply work to project past place into present and future place. This process is fractured—ruptured by different narratives that jostle for attention. Most places have stories that achieve some kind of preeminence. They also have stories that become hidden or silenced only to abruptly reemerge as new constellations of place appear.

> It's a good place to buy stolen hubcaps, stolen auto parts, stolen tires. It must be the Midwest's biggest outlet of these items....
>
> If you live in the city, especially near UIC, and your hubcaps disappear or your car sound system is ripped out, rush to Maxwell Street and you might be able to buy them back....
>
> So if I have to take sides between the expansion of UIC, the city's first public university, which has provided thousands of working class kids with good educations, or a faded and fading Maxwell Street, I'll take the school.
>
> The future of our society is not to be found in hot hubcaps.
>
> **MIKE ROYKO**[83]

> For time genuinely to be held open, space could be imagined as the sphere of the existence of multiplicity, or the possibility of the existence of difference. Such a space is the sphere in which distinct stories co-exist, meet up, affect each other, come into conflict or cooperate. This space is not static, not a cross-section through time; it is disrupted, active and generative. It is not a closed system; it is constantly, as space-time, being made.
>
> **DOREEN MASSEY**[84]

Materialities, meanings, and practices from both within and without are persistently gathered in place. The gathering, or weaving, function of place ensures forms of persistence that link place in the past to place in the present.

> Layers of investment patterns, forms of labor, gender roles, migratory streams, and architectural styles: these among many other possibilities accumulate in sites over time. Often they do so in constitution with one another, with other places, and across geographical scales. Each round of sedimentation cannot help but shape the subsequent round. The place that materializes from this repetitive superimposition is never finished, never closed, never determined. Rather, places understood this way are processual, porous, and articulated.
>
> **PATRICIA PRICE**[85]

Practice/ Temporality

Practice appears to have the most tenuous hold on persistence. The individual doings of people and things (or people with things) are often seen as the site for irreducible singularity.

> Performance cannot be saved, recorded, documented, or otherwise participate in the circulation of representations of representation. . . . Performance's being . . . becomes itself through disappearance.
>
> **PEGGY PHELAN**[86]

In a number of ways, the things we do contribute to the persistence of place. Many theories of practice are centered on the notion of iteration and reiteration. Gender, for instance, has been seen as a product of iterative practice—constantly produced by repetition of actions.[87] Social hierarchy more generally has been theorized as a product of practice and performativity.[88] This focus has highlighted the *becoming* nature, and tenuous hold, of things (gender, society) that appeared to have been already achieved. If people and things are practiced or performed differently, then these seemingly monolithic things would be transformed.

> Performance also constitutes the methodological lens that enables scholars to analyze events *as* performance. Civic disobedience, resistance, citizenship, gender, ethnicity, and sexual identity, for example, are rehearsed and performed daily in the public sphere. To understand these *as* performance suggests that performance also functions as an epistemology. Embodied practice, along with and bound up with other cultural practices, offers a way of knowing.
>
> **DIANA TAYLOR**[89]

Practice is linked to temporality. While an individual act is doomed to disappear as soon as it happens, practices accumulate and repeat. This passage of practice through time links it to the production of meaning.

> First, to recall, recount, or reactivate a scenario we need to conjure up the physical location (the "scene" as physical environment, such as a stage or place in English; *escenario*, a false cognate, means stage in Spanish). *Scene* denotes intentionality, artistic or otherwise (the scene of the crime), and

signals conscious strategies of display. The word appropriately suggests both the material stage as well as the highly codified environment that gives viewers pertinent information, say, class status or historical period. The furnishings, clothing, sounds, and style contribute to the viewer's understanding of what might conceivably transpire there. The two, scene and scenario, stand in metonymic relationship: the place allows us to think about the possibilities of the action. But action also defined place. If, as de Certeau suggests, "space is a practiced space," then there is no such thing as place, for no place is free of history and social practice.
DIANA TAYLOR[90]

Habitual practice leaves imprints on the materiality of place. Architects, planners, and poets refer to the lines eroded into vegetation by people and other animals as "desire lines."

DESIRE LINES (4)

Up on the common
where the bracken's thick
they criss-cross the land
revealing the settled
will of sheep.

One, somewhere
walked that line first
a maverick
in terra incognita

and then the first follower

There are a number of ways in which memory and place are implicated in the production of each other. One technology of memory is the archive. The archive rests on the collection of things. It has a material presence that endures. That is one point of an archive—to ensure endurance. Places can be thought of as archives—as collections of things with origins in the past, things invested with different interpretive possibilities and that continue to be interpreted by people with different aims in mind.

People become obsessed with material remnants because *the past* is a fiction: what remains are memories that are defined by our mourning for that which can no longer be present. We try to preserve memory by creating traces of a past that by definition can never be present. When places are made and understood in this way, their perceived material or emotive presence may seem comforting in the present moment because they are interpreted as giving the past a material form.
KAREN TILL[91]

Repertoire is another technology of memory and stands in contrast to the archive. The repertoire is centered on the importance of performance to memory and focuses on processes of reiteration. Again, place plays an important role. Places are both the settings for repertoires of repeated practice (providing the scenes in which scenarios are played out) and the products of the repertoire's enactments.

> Certainly it is true that individual instances of performances disappear from the repertoire. This happens to a lesser degree than the archive. The question of disappearance in relation to the archive and the repertoire differs in kind as well as degree. The live performance can never be captured or transmitted through the archive. . . . Embodied memory, because it is live, exceeds the archive's ability to capture it. But this does not mean that performance—as ritualized, formalized, or reiterative behavior—disappears. Performances also replicate themselves through their own structures and codes. . . . Multiple forms of embodied acts are always present, though in a constant state of againness. They reconstitute themselves, transmitting communal memories, histories, and values from one group/generation to the next. Embodied and performed acts generate, record, and transmit knowledge.
> **DIANA TAYLOR**[92]

> Buildings are not what makes Maxwell Street historic; street vendors, musicians, and ambience did and [do]. The complex bringing together of persons from all walks of life—poor, rich, black, white, young, old, Hispanic, Jewish, eccentric, intellectual—is what makes Maxwell Street historic. The continuous use of this same place . . . as a poor person's business incubator for 125 years is what makes Maxwell Street historic.
> **WILLIAM GARFIELD, 1994**[93]

There are clear lines to be drawn between the here-ness of place, ways in which places are constituted relationally, and the passage of place through time. What gets lost in a relational approach to place is the specificity of here-ness that resists or attracts particular kinds of flows and relations with the outside. I want to balance the vertical and the horizontal. The suggestion that places are made relationally sets up a false argument of territory and region as static concepts that need to be replaced. In fact the inhabitants of place continually assume space to be formed territorially and act as though it is. These territories are being made and remade all the time through political, economic, and cultural practices. They are, like gender, products of iterative processes. Similarly, networks, the favored spatial form of many relational thinkers, are not always fluid and dynamic but have their own fairly static components or nodes, routes, and moorings.

> Sociospatial relations . . . are deeply processual and practical outcomes of strategic initiatives undertaken by a wide range of forces produced neither through structural determinism nor through a spontaneous voluntarism, but

> through a mutually transformative evolution of inherited spatial structures and emergent spatial strategies within an actively differentiated, continually evolving grid of institutions, territories and regulatory activities. . . . In short, constructed and always emergent space matters in shaping future trajectories.
>
> **MARTIN JONES**[94]

What is it that is being related in relational approaches to place? How come some things—some places—are rich in relations while others are relationally impoverished? Who gets to make things relate? Is anything nonrelational? Might there even be structural reasons why some things enter into relations and others do not?

At the core of this approach is the necessity of taking temporality into account when thinking through the interrelations between relational space and forms of territory and region—the interrelations between fixity and flow in place.

> The assemblage of buildings, land-use patterns, and arteries of communication that constitute place as a visible scene cannot emerge fully formed out of nothingness and stop, grow rigid, indelibly etched in the once-natural landscape. Whether place refers to a village or a metropolis, an agricultural area or an urban-industrial complex, it always represents a human product. Place, in other words, always involves an appropriation and transformation of space and nature that is inseparable from the reproduction and transformation of society in time and space. As such, place is not only what is fleetingly observed on the landscape, a locale, or setting for activity and social interaction. . . . It also is what takes place ceaselessly, what contributes to history in a specific context through the creation and utilization of a physical setting.
>
> **ALLAN PRED**[95]

Martin Jones uses the notion of "phase space," which he borrows from physics (via assemblage theory). Phase space describes the way one set of spatial possibilities leads (or does not lead) to a new arrangement of space at some future point.

> It involves conceding that there *may* be certain circumstances in which, as an object of analysis, practical and bounded spaces that have been institutionalized through particular struggles and become identified as discrete territories in the spheres of economics, politics and culture, *matter*.
>
> **MARTIN JONES**[96]

Space, Jones argues, is "sticky," and things can get stuck or anchored in it. This shapes the unfolding of space and society over time through a familiar interaction of structure (institutions, imperatives, etc.) and agency (individual acts). This stickiness can be thought of as the particularity of places out of which histories unfold.

The notion of the structure of a space of possibilities is crucial in assemblage theory given that, unlike properties, the capacities of an assemblage are not given, that is, they are merely possible when not exercised. But the set of possible capacities of an assemblage is not amorphous, however open-ended it may be, since different assemblages exhibit different sets of capacities.
MANUEL DELANDA[97]

Gathering gives to place its peculiar enduringness, allowing us to return to it again and again as the same place and not just as the same position or site. . . . A place is generative and regenerative on its own schedule. From it experiences are born and to it human beings return for empowerment. . . . A place is more an event than a thing to be assimilated to known categories.
EDWARD CASEY[98]

Particular constellations of place contain within them a multitude of possibilities for futures—different versions of what happens next. The also reintroduce a notion of boundedness and territory in a terrain that relational theorists are attempting to dissolve completely. But territoriality is not the only process at play in the development of place through time.

As gatherings of materialities, meanings, and practices, places are projected forward in heterogeneous ways. Each of these elements of place contributes to the obduracy, or otherwise, of place—as persistent landscape, as memory, as repertoire.

Power

Places do not emerge organically. They are not trees or even rhizomes. Gathering, weaving, and assembling are all subject to, and productive of, the influence of power. How materialities, meanings, and practices are assembled in a particular location is a political process.

I have defined the concept of power by saying that A exercises power over B when A affects B contrary to B's interests.
STEVEN LUKES[99]

Perhaps the most commonsense version of power is the capacity to control the actions of others. Related to this is the idea of power as something that is *used* to achieve a desired outcome—not unlike the way we might use electricity as power. Finally, there is power as its own kind of assemblage, an assemblage of words and practices that defines what is real, true, and possible—even the nature of one's true interests.

We must cease once and for all to describe the effects of power in negative terms: it "excludes," it "represses," it "censors," it "abstracts," it "masks," it "conceals." In fact power produces; it produces reality; it produces domains of objects and rituals of truth. The individual and the knowledge that may be gained of him belong to this production.
MICHEL FOUCAULT[100]

Power takes many forms. Geographer John Allen lists versions of power including domination, authority, seduction, manipulation, and coercion. All appear to refer to ways of affecting an outcome in others, but each has subtly different implications.

> Power, as I understand it, is a relational effect of social interaction. It may bridge the gap between here and there, but only through a succession of mediated relations or through the establishment of a simultaneous presence. People are placed by power, but they experience it at first hand through the rhythms and relationships of particular places, not as some pre-packaged force from afar and not as a ubiquitous presence.
>
> **JOHN ALLEN**[101]

Power is not a property of a person, thing, or place but rather the outcome of relations between people, things, and places. Power may be used to affect unwanted change in others, and it may be used collectively to accomplish a shared project. Power exists in and through place.

Power, like any supposed universal force, only comes into being by engaging with the specific. Power works through places. As soon as it does so, it loses its universality and become specific. Power that is *here* and not *there.*

As a *necessary social construct*, place is present in all stages of the conduct of power. We cannot escape place, and yet every place we inhabit could be otherwise. Place is a result of the play of power, a setting for the conduct of power, and an agent in the production of power. As a form of gathering or assemblage, place becomes a syncretic context within which we have no choice but to lead our lives. The form this gathering takes is fundamental to what is possible or impossible. Its hold over us is subtle and, for the most part, below the level of consciousness. Place is an unavoidable part of the "reality" that power produces and at the same time is a key part of the process through which "reality" and "truth" are produced.

Location, locale, and sense of place, the three aspects of place, all have roles in the play of power. *Where* something is is important. Two Mayor Daleys wanted the UIC campus to be on the edge of the downtown Loop—close to the epicenter of their visions for a successful Chicago. The Maxwell Street Market was in the way. This place that has once been the "zone in transition," the "ghetto," the "black belt" was deemed too central to be a place of suburban affluence, not central enough to be part of the central business district.

> Five minutes from the Loop, adjoining the University of Illinois at Chicago campus. The City's most thoughtfully planned new neighborhood . . .
>
> **CHICAGO'S NEXT GREAT NEIGHBORHOOD**[102]

Location/ Power

Location is more than just where. The definition of spatial extent and the marking of boundaries locate a place that has *area*. This is the elemental beginning of human territoriality as a basic form of power that defines what exists and what is possible.

> At this point let me define what I mean by territoriality explicitly: the attempt by an individual or group (x) to influence, affect, or control objects, people, and relationships (y) by delimiting and asserting control over a geographic area. This area is the territory.
> **ROBERT DAVID SACK**[103]

Territory is an aspect of place but not the same thing. An urban renewal area has a territory with a legally codified boundary that acts to define what can and cannot take place within it. Maxwell Street as a place has no such boundaries. It overspills its limits. Nevertheless, the repeated layering of territories—each with different purposes—plays an important role in the repertoires of power. The various territories create the possibility for different futures.

Locale/Power

Locale—place as the material setting for the conduct of social relations—is also important. We have no choice but to engage with place-as-locale as we move through it, inhabit it, conduct our habits and rituals. The material topography of place enables and limits. It gives us cues as to how to behave. Sometimes we accidently or deliberately fail to follow these cues, but most of the time we do. And thus place is reproduced.[104]

> Traditional Chicago architecture on tree-lined streets with neighborhood parks and green space corridors . . .
> **CHICAGO'S NEXT GREAT NEIGHBORHOOD**[105]

Sense of Place/Power

Sense of place—the individual and shared meanings ascribed to place—is implicated in the choreography of power and place. "Place branding" is the most obvious and instrumental form of producing meaning for place in order to increase its economic value.[106] The labeling of the Maxwell Street area as "University Village," and later, "University Village Maxwell Street," is a clear example of this strategy. All the sedimented layers of description of the market as derelict waste-space perform the opposite, a diminution of value, in a less orchestrated and instrumental way. The possibility of the former depended, to a large degree, on the existence of the latter.

> The mix of dining, retail and service offerings on the south end of the University of Illinois at Chicago campus has been renamed University Village Maxwell Street. The new name reflects the location's historic beginnings dating back to the mid-1800s when the legendary Maxwell Street open air market served as the gateway for immigrants from around the world.
>
> Spared from the Great Chicago Fire of 1871 that started on its doorsteps, the

area went on to become the birthplace of the Chicago Blues, Maxwell Street Polish sausages, musician Benny Goodman's career, and more.

Today, University Village Maxwell Street is a family-friendly destination with dozens of restaurants, taverns, retail outlets, and service shops all located along Halsted and Maxwell Streets, Roosevelt Road and Union Avenue.

PRWEB.COM[107]

Inside/ Outside/ Power

In addition to the locational and territorial definition of place, there is the relation between what is inside the place and what is outside it. A key question is how places enroll or forbid connections with the world beyond. This aspect of place exists hand in hand with territoriality. Mapping and defining the Maxwell Street area as a TIF district enabled particular sets of connections with other places—places where tax increment financing had originated, for instance, or places where workers held pensions with funds invested in the bonds used to finance TIF development. In its earlier history, the existence of Maxwell Street Market as a place of heterogeneity and immigration connected it through migration to Eastern Europe, to the American South, and to Central America.

If Maxwell Street had been declared a historic place and listed on the National Register it would have been connected to a *different* network of materiality, meaning, and practice. It would have been connected to Ellis Island, Beale Street in Memphis, and the battlefields of Gettysburg. A different form of productive power would have been mobilized. All the expertise of the preservation movement would have been enlisted into its assemblage, and it would have become a different place. Relationality does not just happen, it is struggled over.

At the same time that Maxwell Street was enrolling and internalizing its connections, it was also making them possible. Concepts such as "capital" or "urban renewal," which appear to float over the surface of the earth, need place to exist.

> A study of global connections shows the grip of encounter: friction. A wheel turns because of its encounter with the surface of the road; spinning in the air it goes nowhere. Rubbing two sticks together produces heat and light; one stick alone is just a stick. As a metaphorical image, friction reminds us that heterogeneous and unequal encounters can lead to new arrangements of culture and power.
>
> **ANNA LOWENHAUPT TSING**[108]

Just as places are produced through connections, so connections are produced through places.

One aspect of the transformation of the Maxwell Street area was the reconfiguring of its web of connections. This relational aspect of place is also a product and producer of power. The territorial aspects of place at Maxwell Street are not opposed to but, rather, productive of the relational aspects of place. There has been constant struggle over the networks of relationality into which Maxwell Street has been enrolled. Territorial definition has enabled entry into networks of urban renewal and tax increment financing. Earlier senses of place were both productive of, and a result of, global networks of migration and immigration. Maxwell Street has always been both/and—particular and connected.

The assemblage of materialities, meanings, and practices that constitute place makes place a privileged locus of power. Power as a productive force involves the definition of the terms of reality—the conditions under which truth can be assessed. The constant linking of materiality with subjective worlds and domains of reiterative practice in place means that place is one of the primary ways in which power is constituted.

While a command, or a gun pointed at the head, represent obvious examples of the exercise of power, the power of place is more profound and encompassing. The places we live in define the realm of possibility through their materiality (we can go there but not here), their concretization of meaning (this is a church so you shall show reverence), and their role as *scenarios* of repertoires of practice (the people in the library are silent so I will be silent too).

Place appears as second nature—as "what is"—and is thus particularly powerful in the way it connects worlds of materiality, worlds of meaning, and worlds of practice.

> The fish don't talk about the water
> **SRI LANKAN PROVERB**

An account of the links between place and power is not (only) an account of place as a weapon in the reproduction of capitalism or place as a tool for the replication of patriarchy, nor is it a diagram of humanity's domination of the natural world. Places can certainly be those things, but such accounts would fail to complete our understanding of the fundamental importance of place as the very water in which we swim. Place as an assemblage of materiality, meaning, and practice; place as a necessary social construct; place as a crossroads of roots and routes, where pasts become futures—this is where the possibilities for certain kinds of truth are formed, and this is why they are powerful.

Notes

PART ONE

1. Municipal Code of Chicago, quoted in Institute of Urban Life, Loyola University, "Diagnostic Survey of Relocation Problems of Non-Residential Establishments, Roosevelt-Halsted Area" (prepared for the Department of Urban Renewal, City of Chicago, 1965), 21.
2. Louis Wirth, *The Ghetto* (Chicago: University of Chicago Press, 1928), 231–32.
3. David Kishik, *The Manhattan Project: A Theory of a City* (Palo Alto, CA: Stanford University Press, 2015), 21.
4. Jane Rendell, *Site-Writing: The Architecture of Art Criticism* (London: I. B. Tauris, 2010), 2.
5. Theodor Adorno, letter to Walter Benjamin, in Adorno and Benjamin, *Complete Correspondence 1928–1940* (London: Polity, 1999), 283.
6. John Fraser Hart, "The Highest Form of the Geographer's Art," *Annals of the Association of American Geographers* 72, no. 1 (1982).
7. H. C. Darby, "The Problem of Geographical Description," *Transactions of the Institute of British Geographers* 30 (1962): 2.
8. Clifford Geertz, *The Interpretation of Cultures* (New York: Basic Books, 1973), 16.
9. Geertz, *Interpretation of Cultures*, 26.
10. Elizabeth Bishop, letter to Marianne Moore, quoted in Lorrie Goldensohn, *Elizabeth Bishop: The Biography of a Poetry* (New York: Columbia University Press, 1992), 104.
11. Georges Perec, *Species of Spaces and Other Pieces* (New York: Penguin, 1997), 50–51.
12. Perec, *Species of Spaces*, 53.
13. See, for instance, Laura Ogden, *Swamplife: People, Gators, and Mangroves Entangled in the Everglades* (Minneapolis: University of Minnesota Press, 2011); Patricia L. Price, *Dry Place: Landscapes of Belonging and Exclusion* (Minneapolis: University of Minnesota Press, 2004); Kathleen Stewart, *Ordinary Affects* (Durham, NC: Duke University Press, 2007); David Matless, *In the Nature of Landscape: Cultural Geography on the Norfolk Broads* (Oxford: Wiley Blackwell, 2014); Kishik, *Manhattan Project*.
14. Matless, *In the Nature of Landscape*, 25.
15. Roland Barthes, *The Semiotic Challenge* (Berkeley: University of California Press, 1988), 187.

16. Gene Morgan, "Maxwell Street to Have Face-Lifting Operation," *Chicago Daily News*, May 24, 1939, 1.
17. Stanley Fish, *How to Write a Sentence* (New York: HarperCollins, 2011), 84.
18. William Least Heat-Moon, *Prairyerth (a Deep Map)* (Boston: Houghton Mifflin, 1991), 327.
19. Roland Barthes, *A Lover's Discourse* (New York: Hill & Wang, 1978), 7.
20. Anna Lowenhaupt Tsing, *The Mushroom at the End of the World: On the Possibility of Life in Capitalist Ruins* (Princeton, NJ: Princeton University Press, 2015), viii.
21. Kathleen Stewart, *A Space on the Side of the Road* (Princeton, NJ: Princeton University Press, 1996), 7.
22. Claudia Rankine, *Citizen: An American Lyric* (Minneapolis, MN: Graywolf, 2014); Susan Howe, *Pierce-Arrow* (New York: New Directions, 1999); Maggie Nelson, *Jane: A Murder* (Brooklyn, NY: Soft Skull, 2005); Nelson, *Bluets* (Seattle: Wave Books, 2009); Nelson, *The Argonauts* (Seattle: Wave Books, 2015).
23. Nelson, *Argonauts*, 3.
24. Chris Ware, "Chris Ware on *Here* by Richard McGuire—a Game-Changing Graphic Novel" *Guardian*, December 17, 2014, https://www.theguardian.com/books/2014/dec/17/chris-ware-here-richard-mcguire-review-graphic-novel.
25. Chris Ware, *Building Stories* (New York: Pantheon, 2012); Richard McGuire, *Here* (New York: Pantheon, 2014).
26. Walter Benjamin, *The Arcades Project* (Cambridge, MA: Belknap Press of Harvard University Press, 1999); Heat-Moon, *Prairyerth*; Bruce Chatwin, *The Songlines* (Harmondsworth: Penguin, 1988).
27. Allan Pred, "Hypermodernity, Identity and the Montage Form," in *Space and Social Theory: Interpreting Modernity and Postmodernity*, ed. Georges Benko and Ulf Strohmayer (Oxford: Blackwell, 1997), 135.
28. Benjamin, *Arcades Project*, 460.
29. Kenneth Goldsmith, *Capital: New York, Capital of the 20th Century* (New York: Verso, 2015).
30. Kenneth Goldsmith, "Rewriting Walter Benjamin's 'The Arcades Project,'" http://www.poetryfoundation.org/harriet/2011/04/rewriting-walter-benjamins-the-arcades-project (accessed December 10, 2017).
31. William Cronon, "A Place for Stories: Nature, History, and Narrative," *Journal of American History* 78, no. 4 (1992): 1349.
32. Hayden White, "The Question of Narrative in Contemporary Historical Theory," *History and Theory* 23, no. 1 (1984): 1.
33. Robert Darnton, "Extraordinary Commonplaces," *New York Review of Books*, December 21, 2000, 82.
34. Tim Cresswell, *Soil* (London: Penned in the Margins, 2013); Cresswell, *Fence* (London: Penned in the Margins, 2015); Sarah de Leeuw, *Geographies of a Lover* (Edmonton: NeWest Press, 2012); de Leeuw, *Skeena* (Halfmoon Bay, BC: Caitlin Press, 2015); Eric Magrane and Christopher Cokinos, eds., *The Sonoran Desert: A Literary Field Guide* (Tucson: University of Arizona Press, 2016); Nicholas Bauch, *Enchanting the Desert* (Palo Alto, CA: Stanford University Press, 2016). For a collection of creative geographies, see "Creativity and Geography," ed. Sarah de Leeuw and Sallie Marston, a special issue of *Geographical Review* 103, no. 2 (2013). For wider accounts of creative geographies, see Harriet Hawkins, *For Creative Geographies: Geography, Visual Arts and the Making of Worlds* (London: Routledge, 2013).

35. Donald Meinig, "Geography as an Art," *Transactions of the Institute of British Geographers* 8 (1983): 325.
36. But see Simon Springer, "Earth Writing," *GeoHumanities* 3, no. 1 (2017).
37. For more on this, see Tim Cresswell, "Towards Topopoetics: Space, Place and the Poem," in *Place, Space and Hermeneutics*, ed. Bruce Janz (New York: Springer, 2017), 319–31.
38. For a definition, see "Aristotle's Rhetoric," *Stanford Encyclopedia of Philosophy*, http://plato.stanford.edu/entries/aristotle-rhetoric/#7.1.
39. Aristotle, *Topics*, in *The Complete Works of Aristotle: The Revised Oxford Translation*, vol. 1 (Princeton, NJ: Princeton University Press, 1984), 163b28.32.
40. Jeff Malpas, *Heidegger's Topology: Being, Place, World* (Cambridge, MA: MIT Press, 2008), 69.
41. Stanley Fish, "Barack Obama's Prose Style" *New York Times*, January 22, 2009, https://fish.blogs.nytimes.com/2009/01/22/barack-obamas-prose-style/?scp=3&sq=stanley%20fish&st=cse.
42. Susan Buck-Morss, *The Dialectics of Seeing: Walter Benjamin and the Arcades Project* (Cambridge, MA: MIT Press, 1991), 336.
43. Barthes, *Lover's Discourse*, 9.
44. Rebecca Solnit, *A Field Guide to Getting Lost* (London: Penguin, 2006), 6.
45. Simone de Beauvoir, "America Day by Day (L'amerique au jour le jour)," in *Building the Nation: Americans Write about Their Architecture, Their Cities, and Their Landscape*, ed. Steven Conn and Max Page (Philadelphia: University of Pennsylvania Press, 2003), 69.
46. Kathleen Stewart, "Precarity's Forms," *Cultural Anthropology* 27, no. 3 (2012): 518.
47. Nelson, *Bluets*, 91.
48. Bob Perelman, "Parataxis and Narrative: The New Sentence in Theory and Practice" *American Literature* 65, no. 2 (1993).
49. William Cronon, *Nature's Metropolis: Chicago and the Great West* (New York: Norton, 1991), 23.
50. William Cronon, "Kennecott Journey: The Paths Out of Town," in *Under an Open Sky*, ed. Cronon, George Miles, and Jay Gitlin (New York: Norton, 1992).
51. "Death of Mr. Philip Maxwell," *Chicago Press & Tribune* 13, no. 3 (November 8, 1859).
52. Wirth, *Ghetto*, 195.
53. Cronon, *Nature's Metropolis*, 5.
54. James Burkhart Gilbert, *Perfect Cities: Chicago's Utopias of 1893* (Chicago: University of Chicago Press, 1991).
55. Alan P. Mamoser, "Requiem for Maxwell Street," *Newcity*, December 28, 2000, 8.
56. Cronon, "Place for Stories," 1354.

PART TWO

1. "Chicago's Next Great Neighborhood" (brochure), UIC Archives, File RR 44.2.
2. Willard Motley, *Let No Man Write My Epitaph* (New York: Random House, 1958), 92.
3. Curtis Lawrence, "City Dealing to Preserve Maxwell Street," *Chicago Sun-Times*, December 3, 1999.
4. Ira Berkow, *Maxwell Street: Survival in a Bazaar* (Garden City, NY: Doubleday, 1977); Shuli Eshel and Roger Schatz, *Jewish Maxwell Street Stories* (Charleston, SC: Arcadia, 2004).

5. Carl Sandburg, "Fish Crier," from *Chicago Poems* (1916), http://carl-sandburg.com/fish_crier.htm (accessed February 11, 2018).
6. Greg Trotter, "Pre-Chicago Fire Building Being Demolished in Maxwell Street Area," *Chicago Tribune*, March 7, 2015.
7. Charles Zeublin, "The Chicago Ghetto," in *Hull House Maps and Papers*, ed. Jane Addams (New York: Thomas Y. Crowell & Co., 1895), 92.
8. Clarence Darrow, "Little Louis Epstine" (1903), in *The Essential Words and Writings of Clarence Darrow* (New York: Random House, 2007), 172.
9. Hilda Polacheck, "The Ghetto Market" (1905), quoted in *I Came a Stranger: The Story of a Hull-House Girl* (Urbana-Champaign: University of Illinois Press, 1991); 78.
10. De Beauvoir, "America Day by Day," 69.
11. Michael Raleigh, *The Maxwell Street Blues: A Chicago Mystery Featuring Paul Whelan* (New York: St. Martin's Press, 1994); Alfonso Morales, "Making Money at the Market: The Social and Economic Logic of Informal Markets" (PhD diss., Northwestern University, 1993); Berkow, *Maxwell Street: Survival in a Bazaar*; Lori Grove, *Chicago's Maxwell Street* (Chicago: Aracadia, 2002).
12. Edward C. Schulz, "A Functional Analysis of Retail Trade in the Maxwell Street Market Area of Chicago" (master's thesis, Northwestern University, 1954).
13. Jack Lait and Lee Mortimer, *Chicago: Confidential!* (New York: Crown, 1950); *New York: Confidential!* (Chicago: Ziff-Davis, 1948); *Washington: Confidential!* (New York: Crown, 1951).
14. Will Straw, "Urban Confidential: The Lurid City of the 1950s," in *The Cinematic City*, ed. David Clarke (London: Routledge, 1997).
15. Lait and Mortimer, *Chicago: Confidential!*, 60.
16. Lait and Mortimer, *Chicago: Confidential!*, 60–61.
17. Chad Heap, *Slumming: Sexual and Racial Encounters in American Nightlife, 1885–1940* (Chicago: University of Chicago Press, 2009), 2.
18. "Maxwell Street Heart of Ghetto," *Chicago Daily News*, April 28, 1928, 14.
19. Stephen Morris, "Maxwell Street Update." *Chicago Tribune*, October 30, 1981, C3.
20. Louise Crewe, "Life Itemised: Lists, Loss, Unexpected Significance, and the Enduring Geographies of Discard," *Environment and Planning D: Society and Space* 29, no. 1 (2011): 30.
21. Umberto Eco, *The Infinity of Lists* (London: MacLehose, 2009).
22. Georges Perec, *An Attempt at Exhausting a Place in Paris* (Cambridge, MA: Wakefield Press, 2010), 11–12.
23. Michel Foucault, *The Order of Things: An Archaeology of the Human Sciences* (New York: Pantheon, 1971), xv.
24. Borges presents this taxonomy, purportedly from an ancient Chinese text, in his essay "The Analytical Language of John Wilkins" (1942), reprinted in *Selected Nonfictions*, trans. Eliot Weinberger (New York: Penguin, 1999).
25. Foucault, *Order of Things*, xvi.
26. Yi-Fu Tuan, *Space and Place: The Perspective of Experience* (Minneapolis: University of Minnesota Press, 1977), 178.
27. Paul Rodaway, *Sensuous Geographies: Body, Sense, and Place* (London: Routledge, 1994).
28. Ros Bandt, Michelle Duffy, and Dolly MacKinnon, *Hearing Places: Sound, Place, Time and Culture* (Newcastle: Cambridge Scholars, 2009); Steven Feld and D. Brenneis, "Doing Anthropology in Sound," *American Ethnologist* 31, no. 4 (2004).
29. Tuan, *Space and Place*, 16.

30. Steven Feld, "Waterfalls of Song: An Acoustemology of Place Resounding in Bosavi, Papua New Guinea," in *Senses of Place*, ed. Feld and Keith H. Basso (Santa Fe, NM: School of American Research Press, 1996), 97.
31. Michael Haberlandt, *Cultur im Alltag: Gesammelte Aufsätze von Michael Haberlandt* (1900), 178, quoted in Karin Bijstervald, *Mechanical Sound: Technology, Culture, and Public Problems of Noise in the Twentieth Century* (Cambridge, MA: MIT Press, 2008), 13.
32. Lloyd Wendt, "Business Is Always Good on Maxwell Street," *Chicago Sunday Tribune*, October 9, 1947.
33. Oscar Katov, "Maxwell St. Has Air of Baghdad Bazaars," *Chicago Sun Times*, May 24, 1951.
34. Zygmunt Bauman, *Wasted Lives: Modernity and Its Outcasts* (Oxford: Polity, 2003), 22.
35. "Maxwell Street Heart of Ghetto," *Chicago Daily News*, April 28, 1928, 14.
36. Abe Martin, "Memories of Maxwell Street," *Chicago Tribune*, September 17, 1957.
37. Angela Parker, "Flea Market Spirit Lives on Maxwell Street," *Chicago Tribune*, September 20, 1970.
38. Roger Simon, "Capturing Song of Maxwell St.," *Chicago Sun-Times*, September 10, 1977.
39. David Whiteis, "The Sunday Morning Market May Be in Danger, but Thanks to a New Generation of Bluesmen the Music Is as Strong as Ever," *Chicago Reader*, October 14, 1988, 8.
40. "Maxwell Street Blues," *Chicago Sun-Times*, December 24, 1989, 1.
41. Susanne Hauser, "Waste into Heritage: Remarks on Materials in the Arts, on Memories and the Museum," in *Waste Site Stories: The Recycling of Memory*, ed. Brian Neville and Johanne Villeneuve (Albany, NY: SUNY Press, 2002), 40.
42. Mary Douglas, *Purity and Danger: An Analysis of Concepts of Pollution and Taboo* (New York: Praeger, 1966).
43. Alain Corbin, *The Foul and the Fragrant: Odor and the French Social Imagination* (Cambridge, MA: Harvard University Press, 1986); William Ian Miller, *The Anatomy of Disgust* (Cambridge, MA: Harvard University Press, 1997).
44. David Trotter, "The New Historicism and the Psychopathology of Everyday Modern Life," in *Filth: Dirt, Disgust, and Modern Life* (Minneapolis: University of Minnesota Press, 2005), 38.
45. Trotter, "New Historicism," 39.
46. Nigel Thrift, "All Nose," in *Handbook of Cultural Geography*, ed. Kay Anderson et al. (London: Sage, 2002), 10.
47. Bauman, *Wasted Lives*.
48. Igor Kopytoff, "The Cultural Biography of Things: Commoditization as Process," in *The Social Life of Things: Commodities in Cultural Perspective*, ed. Arjun Appadurai (Cambridge: Cambridge University Press, 1986).
49. James R. Giles and Jerome Klinkowitz, "The Emergence of Willard Motley in Black American Literature," *Negro American Literature Forum* 6, no. 2 (1972).
50. Willard Motley, *Knock on Any Door* (New York: Appleton-Century, 1947).
51. Essentially a street thief.
52. Motley, *Let No Man Write My Epitaph*.
53. Robert Bone, "Richard Wright and the Chicago Renaissance," *Callaloo* 28 (1986).
54. Willard Motley Papers, Northern Illinois University Library, Special Collections, box 13, folder 42.
55. R. Park and E. Burgess, *The City: Suggestions for Investigation of Human Behavior in the Urban Environment* (Chicago: University of Chicago Press,

1925); Harvey Warren Zorbaugh, *The Gold Coast and the Slum: A Sociological Study of Chicago's Near North Side* (Chicago: University of Chicago Press, 1929); Ulf Hannerz, *Exploring the City: Inquiries toward an Urban Anthropology* (New York: Columbia University Press, 1980); Carla Cappetti, *Writing Chicago: Modernism, Ethnography and the Novel* (New York: Columbia University Press, 1993).

56. Ernest Burgess, "The Growth of the City: An Introduction to a Research Project," in Park and Burgess, *The City*.

57. This can be found in Willard Motley Papers, Northern Illinois University Library, Special Collections, box 68, file 6.

58. Bone, "Richard Wright and the Chicago Renaissance"; Cappetti, *Writing Chicago*.

59. Clifford R. Shaw and E. W. Burgess, *The Jack-Roller: A Delinquent Boy's Own Story* (Chicago: University of Chicago Press, 1930).

60. Willard Motley and Jerome Klinkowitz, *The Diaries of Willard Motley* (Ames: Iowa State University Press, 1979), 165.

61. Motley and Klinkowitz, *Diaries*, 179.

62. Willard Motley Papers, Northern Illinois University Library, Special Collections, box 13, folder 44.

63. Willard Motley Papers, Northern Illinois University Library, Special Collections, box 6, folder 2, diary 1939.

64. Michel Serres, *The Five Senses: A Philosophy of Mingled Bodies* (London: Continuum, 2008), 278.

65. Motley, *Knock on Any Door*, 83–84.

66. Willard Motley Papers, Northern Illinois University Library, Special Collections, box 13, folder 42.

67. Willard Motley Papers, Northern Illinois University Library, Special Collections, box 13, folder 42.

68. Motley, *Let No Man Write My Epitaph*, 92–93.

69. See David Seamon, "Body-Subject, Time-Space Routines, and Place-Ballets," in *The Human Experience of Space and Place*, ed. Anne Buttimer and David Seamon (London: Croom Helm, 1980); Allan Pred, "Place as Historically Contingent Process: Structuration and the Time-Geography of Becoming Places," *Annals of the Association of American Geographers* 74, no. 2 (1984); Michel de Certeau, *The Practice of Everyday Life* (Berkeley: University of California Press, 1984).

70. Willard Motley Papers, Northern Illinois University Library, Special Collections, box 13, folder 42.

71. Willard Motley Papers, Northern Illinois University Library, Special Collections, box 13, folder 42.

72. Raymond Madden, *Being Ethnographic: A Guide to the Theory and Practice of Ethnography* (Thousand Oaks, CA: SAGE Publications, 2010).

73. Peter Stallybrass and Allon White, *The Politics and Poetics of Transgression* (Ithaca, NY: Cornell University Press, 1986).

74. W. E. B. Du Bois, *The Souls of Black Folk* (New York: Dover, 1903), 2.

75. John Berger, *Ways of Seeing* (Harmondsworth, UK: Penguin, 1972), 47.

76. Laura Mulvey, "Visual Pleasure and Narrative Cinema," *Screen* 16, no. 3 (1975).

77. See, for instance, Jonathan Crary, *Suspensions of Perception: Attention, Spectacle, and Modern Culture* (Cambridge, MA: MIT Press, 1999).

78. Michel Foucault, *Discipline and Punish: The Birth of the Prison* (New York: Vintage, 1979).

79. Catherine Nash, "Reclaiming Vision: Looking at Landscape and the Body," *Gender, Place and Culture* 3, no. 2 (1996); Subir Bhaumik, *Counter-Gaze: Media, Migrants, Minorities* (Kolkata: Frontpage, 2010).
80. Frederic Milton Thrasher, *The Gang: A Study of 1,313 Gangs in Chicago* (Chicago: University of Chicago Press, 1936); Zorbaugh, *Gold Coast and the Slum*; Paul Goalby Cressey, *The Taxi-Dance Hall: A Sociological Study in Commercialized Recreation and City Life* (Chicago: University of Chicago Press, 1932); Nels Anderson, *The Hobo* (Chicago: University of Chicago Press, 1925).
81. Tim Cresswell, "The Peninsular of Submerged Hope: Ben Reitman's Social Geography," *Geoforum* 29, no. 2 (1998); David Sibley, "Gender, Science, Politics and Geographies of the City," *Gender, Place and Culture* 2, no. 1 (1995).
82. John Urry, *The Tourist Gaze* (London: Sage, 2001), 3.
83. Peter X. Feng, *Screening Asian Americans* (New Brunswick, NJ: Rutgers University Press, 2002), 22.
84. Willard Motley Papers, Northern Illinois University Library, Special Collections, box 13, folder 42.
85. Motley, *Let No Man Write My Epitaph*, 111–12.
86. For a fascinating account of Maier's life see the documentary movie, *Finding Vivian Maier*.
87. For accounts of Vivian Maier, see Richard Cahan and Michael Williams, *Vivian Maier: Out of the Shadows* (Chicago: CityFiles Press, 2012); John Maloof, *Vivian Maier: Street Photographer* (Brooklyn, NY: PowerHouse Books, 2011).
88. For an account of the unstable boundaries between "street photography" and more official forms of photography in the city, see Joseph Heathcott and Angela Dietz, *Capturing the City: Photographs from the Streets of St. Louis, 1900–1930* (St. Louis: Missouri History Museum Press, 2016).
89. Walter Benjamin, "A Short History of Photography," in *Classic Essays on Photography*, ed. Alan Trachtenberg (New Haven, CT: Leete Island Books, 1980), 215.
90. Moholy-Nagy, quoted in Mary Benedetta, *The Street Markets of London* (London: John Miles, 1936), vii–viii.
91. N. Thrift and J. D. Dewsbury, "Dead Geographies—and How to Make Them Live," *Environment and Planning D: Society and Space* 18, no. 4 (2000).
92. This approach to representation has much in common with the view expressed in Ben Anderson and Paul Harrison, "The Promise of Non-Representational Theories," in *Taking Place: Non-Representational Theories and Geography*, ed. Anderson and Harrison (Farnham: Ashgate, 2010).
93. Jacques Derrida, *Copy, Archive, Signature: A Conversation on Photography* (Stanford, CA: Stanford University Press, 2010), 6.
94. Roland Barthes, *Camera Lucida: Reflections on Photography* (1981; London: Flamingo, 1984).
95. Allan Sekula, "On the Invention of Photographic Meaning," in *Photography in Print: Writings from 1816 to the Present*, ed. Vicki Goldberg (Albuquerque: University of New Mexico Press, 1981), 453.
96. Pierre Bourdieu, *Photography: A Middle-Brow Art* (Polity Press, 1990), 6–7.
97. J. D. Dewsbury, "Language and the Event: The Unthought of Appearing Worlds," in *Taking Place: Non-Representational Theories and Geography*, ed. Ben Anderson and Paul Harrison (Farnham: Ashgate, 2010).
98. Joan M. Schwartz and James M. Ryan, *Picturing Place: Photography and the Geographical Imagination* (London: I. B. Tauris, 2003), 1, 7.
99. Heathcott and Dietz, *Capturing the City*, 27.

100. David Harvey, "The Right to the City," *New Left Review* 11 (2008): 34.
101. Ralph Waldo Emerson, *The Conduct of Life* (Boston: Ticknor and Fields, 1860), 129.
102. Moholy-Nagy, quoted in Benedetta, *Street Markets of London*, 8.
103. Clive Scott, *Street Photography: From Atget to Cartier-Bresson* (London: I. B. Tauris, 2007), 8.
104. For accounts of the "slum" and ghetto, see Peter Geoffrey Hall, *Cities of Tomorrow: An Intellectual History of Urban Planning and Design in the Twentieth Century*, 3rd ed. (Oxford: Blackwell, 2002); Felix Driver, "Moral Geographies: Social Science and the Urban Environment in Mid-Nineteenth Century England," *Transactions of the Institute of British Geographers* 13 (1988); David Ward, *Poverty, Ethnicity and the American City, 1840–1925* (Cambridge: Cambridge University Press, 1989). For a classic account with Maxwell Street at its center, see Wirth, *Ghetto*.
105. Sekula, "On the Invention of Photographic Meaning"; Gillian Rose, "Engendering the Slum: Photography in East London in the 1930s," *Gender Place and Culture* 4, no. 3 (1997); Martha Rosler, "In, around, and Afterthoughts (on Documentary Photography)," in *The Contest of Meaning: Critical Histories of Photography*, ed. Richard Bolton (London: MIT Press, 1989); Tim Cresswell, *The Tramp in America* (London: Reaktion, 2001).
106. The classic account of orientalism is Edward W. Said, *Orientalism* (New York: Vintage, 1979). For an account of some of these processes in the city, see Stallybrass and White, *Politics and Poetics of Transgression*.
107. Henry Mayhew, *Mayhew's London (Selections from London Labour and London Poor)* (London: Spring Books, 1851); Charles Booth, *Life and Labour of the People in London*, 9 vols. (London: Macmillan, 1892).
108. Park and Burgess, *The City*; Zorbaugh, *Gold Coast and the Slum*.
109. Jacob Riis, *How the Other Half Lives* (New York: Charles Scribner's Sons, 1890), 296.
110. J. Thomson and Adolphe Smith, *Street Life in London. By J. Thomson . . . and Adolphe Smith. With Permanent Photographic Illustrations Taken from Life Expressly for This Publication* (London: Sampson Low, Marston, Searle, & Rivington, 1877).
111. Stallybrass and White, *Politics and Poetics of Transgression*.
112. Schwartz and Ryan, *Picturing Place*, 8.
113. For the classic account of the role of the gaze in "disciplining" the poor, see Foucault, *Discipline and Punish*. See also Felix Driver, *Power and Pauperism: The Workhouse System, 1834–1884* (Cambridge: Cambridge University Press, 1993).
114. Jacob A. Riis, *The Making of an American* (New York: Macmillan, 1901), 268.
115. See, for instance, Peter Jackson, "Constructions of Culture, Representations of Race; Edward Curtis's 'Way of Seeing,'" in *Inventing Places: Studies in Cultural Geography*, ed. Kay Anderson and Fay Gale (Melbourne: Longman, 1992); Elizabeth Edwards, *Anthropology and Photography, 1860–1920* (New Haven: Yale University Press, 1992); James R. Ryan, *Picturing Empire: Photography and the Visualization of the British Empire* (London: Reaktion, 1997); Laurel Smith, "Chips Off the Old Ice Block: *Nanook of the North* and the Relocation of Cultural Identity," in *Engaging Film: Geographies of Identity and Mobility*, ed. Tim Cresswell and Deborah Dixon (London: Rowman and Littlefield, 2001).
116. Rose, "Engendering the Slum," 280.
117. Shawn Michelle Smith, *American Archives: Gender, Race, and Class in Visual Culture* (Princeton, NJ: Princeton University Press, 2000), 6.

118. Rosler, "In, around, and Afterthoughts," 308.
119. Scott, *Street Photography*.
120. For discussions (and critiques) of this notion of landscape, see Denis E. Cosgrove, "Prospect, Perspective and the Evolution of the Landscape Idea," *Transactions of the Institute of British Geographers* 10 (1985); Tim Cresswell, "Landscape and the Obliteration of Practice," in *Handbook of Cultural Geography*, ed. Kay Anderson et al. (London: Sage, 2003); P. Merriman et al., "Landscape, Mobility, Practice," *Social & Cultural Geography* 9, no. 2 (2008).
121. De Certeau, *Practice of Everyday Life*, 92.
122. Moholy-Nagy, quoted in Benedetta, *Street Markets of London*, viii.
123. For accounts of street photography, see Colin Westerbeck and Joel Meyerowitz, *Bystander: A History of Street Photography* (Boston: Bulfinch, 2001); Scott, *Street Photography*.
124. Robert Frank, *The Americans: Photographs* (New York: Grove, 1959).
125. Quoted in Douglas A. Harper, *Visual Sociology* (New York: Routledge, 2012), 32.
126. See "Daddy Stovepipe," Blues Trail website, http://www.thebluestrail.com/artists/dstove.htm.
127. "Chicago's Chicken Man," *Chicago Stories* (blog), http://pageoption.wordpress.com/2012/01/23/chicagos-chicken-man/ (accessed December 27, 2016).
128. Colin Ward, *The Child in the City* (London: Architectural Press, 1978).
129. Gill Valentine, "Children Should Be Seen and Not Heard: The Production and Transgression of Adult's Public Space," *Urban Geography* 17, no. 3 (1996); Chris Philo, "The Child in the City," *Journal of Rural Studies* 8, no. 2 (1992).
130. Walter Benjamin, "Old Forgotten Children's Books," in *Walter Benjamin: Selected Writings*, vol. 1, *1913–1926*, ed. Marcus Bullock and Michael W. Jennings (Cambridge, MA: Harvard University Press).
131. For instance, at the Institute of Contemporary Art in Boston in 1979 and the University of Iowa Museum of Art in 1993. See Nathan Lerner and Jo-Ann Conklin, *Nathan Lerner's Maxwell Street*, exh. cat. (University of Iowa Museum of Art, 1993); Steven Prokopoff, *Nathan Lerner: A Photographic Retrospective 1932–1979*, exh. cat. (Boston: Institute of Contemporary Art, 1979).
132. Prokopoff, *Nathan Lerner*.
133. Lerner and Conklin, *Nathan Lerner's Maxwell Street*, 7.
134. Geoff Dyer, *The Ongoing Moment* (London: Little, Brown, 2005).
135. Therese Lichtenstein, "The City in Twilight," in *Twilight Visions: Surrealism and Paris*, ed. Lichtenstein (Berkeley: University of California Press, 2009). 18
136. J. K. Huysmans, in Benjamin, *Arcades Project*, 694.
137. Allan Sekula, "On the Invention of Photographic Meaning," in *Photography in Print: Writings from 1816 to the Present*, ed. Vicki Goldberg (Albuquerque: University of New Mexico Press, 1981), 472.
138. Susan Sontag, *On Photography* (New York: Farrar, Straus and Giroux, 1977), 63.
139. Sontag, *On Photography*, 69.
140. Sontag, *On Photography*, 78.
141. Jane Bennett, *Vibrant Matter: A Political Ecology of Things* (Durham, NC: Duke University Press, 2010), 4.
142. Bennett, *Vibrant Matter*.
143. Bennett, *Vibrant Matter*, 5.
144. Ben Highmore, *Everyday Life and Cultural Theory: An Introduction* (London: Routledge, 2002), 47.
145. Andre Breton, *Nadja* (1928; London: Penguin, 1999), 52.
146. Benjamin, *Arcades Project*, 922.
147. Foucault, *Order of Things*, xvii.

148. Jennifer Tucker, "Eye on the Street Photography in Urban Public Spaces," *Radical History Review*, no. 114 (2012), 9.
149. Tucker, "Eye on the Street Photography," 12.
150. Wirth, *Ghetto*, 232–33.
151. Ira Berkow Archives, UIC Library, Special Collections, "Ira Berkow," 78-16, box 1, file 1-1.
152. Max Weber, *The City* (Heinemann, 1960); Jane Jacobs, *The Economy of Cities* (New York: Cape, 1969).
153. Max Weber, "The Nature of the City," in *Classic Essays on the Culture of Cities*, ed. Richard Sennett (Englewood Cliffs, NJ: Prentice-Hall, 1969), 24.
154. Wirth, *Ghetto*, 233.
155. M. Jamieson, "The Place of Counterfeits in Regimes of Value: An Anthropological Approach," *Journal of the Royal Anthropological Institute* 5, no. 1 (1999); K. Schlosser, "Regimes of Ethical Value? Landscape, Race and Representation in the Canadian Diamond Industry," *Antipode* 45, no. 1 (2013); Arjun Appadurai, "Introduction: Commodities and the Politics of Value," in *The Social Life of Things: Commodities in Cultural Perspective*, ed. Appadurai (Cambridge: Cambridge University Press, 1986).
156. Appadurai, "Introduction"; Kopytoff, "Cultural Biography of Things."
157. Michael Shea (Buy a Tux Formal Wear Superstore, West Roosevelt Avenue), letter to office of Richard M. Daley, October 30, 1993, UIC University Archives, Associate Chancellor, South Campus Development Records, 003/02/02, series 1, box 11, folder 11-90.
158. C. J. Johnson (UIC head gymnastics coach), letter to office of Richard M. Daley, November 3, 1993, UIC University Archives, Associate Chancellor, South Campus Development Records 003/02/02, series 1, box 11, folder 11-93.
159. Ira Berkow Archives, UIC Library, Special Collections, "Tyner White," 78-16, box 1, file 1-8, p. 8.
160. Deanna Isaacs, "The Collected and the Ultimate Collector: Tyner White's Glorious, Homeless Hoard," *Chicago Reader*, April 7, 2005, http://www.chicagoreader.com/chicago/the-collected-and-the-ultimate-collector-tyner-whites-glorious-homeless-hoard/Content?oid=918451 (accessed October 21, 2012).
161. Kari Lyderson, "Burnt Out: Will the New Year's Eve Fire Break the Spirit of Preservationists Fighting to Save What's Left of Maxwell Street?" *Chicago Reader*, January 6, 2000, http://www.chicagoreader.com/chicago/burnt-out/Content?oid=901112 (accessed October 21, 2012).
162. Report on public meeting held by City of Chicago Community Development Commission, October 26, 1993, at YMCA, 1001 West Roosevelt Road, Chicago, UIC University Archives, Associate Chancellor, South Campus Development Records, 003/02/02, series X, box 51, file 432, pp. 118–22.
163. Jan Jagodzinski, ed., *Psychoanalyzing Cinema: A Productive Encounter with Lacan, Deleuze, and Žižek* (New York: Palgrave Macmillan, 2012), 6.
164. Susan Strasser, *Waste and Want: A Social History of Trash* (New York: Holt, 2000), 289.
165. James Baldwin, *Conversations with James Baldwin* (Jackson: University of Mississippi Press, 1989), 42.
166. Samuel Zipp, *Manhattan Projects: The Rise and Fall of Urban Renewal in Cold War New York* (Oxford: Oxford University Press, 2010); Hall, *Cities of Tomorrow*.
167. Jane Jacobs, *The Death and Life of Great American Cities* (New York: Vintage, 1961).

168. Herbert Gans (1963), quoted in Hall, *Cities of Tomorrow*, 233.
169. Peter H. Rossi and Robert A. Dentler, *The Politics of Urban Renewal: The Chicago Findings* (New York: Free Press of Glencoe, 1961).
170. Ta-Nehisi Coates, "The Case for Reparations," *Atlantic*, June 2014, https://www.theatlantic.com/magazine/archive/2014/06/the-case-for-reparations/361631/ (accessed November 5, 2017).
171. See Richard Rothstein, *The Color of Law: A Forgotten History of How Our Government Segregated America* (New York: Liveright, 2017)
172. D. Corner, "The Agency of Mapping," in *Mappings*, ed. Denis Cosgrove (London: Reaktion, 1999); J. B. Harley, "Deconstructing the Map," *Cartographica* 26 (1989); Ward L. Kaiser and Denis Wood, *Seeing through Maps: The Power of Images to Shape Our World View* (Amherst, MA: ODT Inc., 2001).
173. Institute of Urban Life, "Diagnostic Survey," 2.
174. Institute of Urban Life, "Diagnostic Survey," 2.
175. Institute of Urban Life, "Diagnostic Survey," 3.
176. Institute of Urban Life, "Diagnostic Survey," 24.
177. Chicago Department of Urban Renewal, Designation of Slum and Blighted Area—Roosevelt-Halsted (Chicago, 1966), 1.
178. Chicago Department of Urban Renewal, Designation of Slum and Blighted Area.
179. Lawrence Veiller, "Slum Clearance," in *Housing Problems in America* (New York: National Housing Association, 1929), 75.
180. Rachel Weber, "Extracting Value from the City: Neoliberalism and Urban Redevelopment," *Antipode* 34, no. 3 (2002): 529.
181. Peter Jackson, "Social Disorganization and Moral Order in the City," *Transactions of the Institute of British Geographers* 9 (1984); Rolf Lindner, *The Reportage of Urban Culture: Robert Park and the Chicago School* (Cambridge: Cambridge University Press, 1996); Hannerz, *Exploring the City*.
182. Burgess, "Growth of the City," 54.
183. Wendell E. Pritchett, "The 'Public Menace' of Blight: Urban Renewal and the Private Uses of Eminent Domain," *Yale Law and Policy Review* 21, no. 2 (2003): 21.
184. Colin Gordon, "Blighting the Way: Urban Renewal, Economic Development, and the Elusive Definition of Blight," *Fordham Urban Law Journal* 31, no. 2 (2003): 307.
185. Evelyn Mae Kitagawa and Karl E. Taeuber, eds., *Local Community Fact Book: Chicago Metropolitan Area, 1960* (Chicago Community Inventory, University of Chicago, 1963), 70.
186. Charles Gruenberg "Slum Study . . . and Footnotes" *Chicago Daily News*, April 10, 1961.
187. John J. Egan (director, Archdiocese Conservation Council), testimony concerning the Near West Side Urban Renewal Project, before the Chicago City Council Committee on Planning and Housing, September 18, 1961.
188. Florence Scala [a local activist], quoted in Berkow, *Maxwell Street: Survival in a Bazaar*, 517.
189. Department of Urban Renewal, City of Chicago, "Roosevelt-Halsted: Proposals for Renewal" (August 1966), 27.
190. Institute of Urban Life, "Diagnostic Survey," 22, 39.
191. Wirth, *Ghetto*, 198.
192. "A Look to the Future: Strategic Plans for UIC" (1987), UIC Archives, file RR 44.0.
193. Chicago Department of Planning, "The Future of the Maxwell Street Market" (1989), 53, UIC Archives, file RR 44.0.

194. Johnson, Johnson and Roy, "Master Plan for the South Campus Development," UIC Library, Special Collections, UIC Archives, file RR44.1.
195. David Whiteis, "The Sunday Morning Market May Be in Danger, but Thanks to a New Generation of Bluesmen the Music Is as Strong as Ever," *Chicago Reader*, October 14, 1988, 8.
196. "Chicago's Next Great Neighborhood," UIC Archives, file RR 44.2.
197. David Harvey, "Neo-Liberalism as Creative Destruction," *Geografiska Annaler Series B: Human Geography* 88B, no. 2 (2006); Harvey, *The Limits to Capital* (Oxford: Blackwell, 1982).
198. Harvey, "Neo-Liberalism as Creative Destruction," 155.
199. Marshall Berman, *All That Is Solid Melts into Air: The Experience of Modernity*, 2nd ed. (Harmondsworth: Penguin, 1988), 99.
200. Harvey, *Limits to Capital.*
201. Weber, "Extracting Value from the City," 519.
202. S. McGreal et al., "Tax-Based Mechanisms in Urban Regeneration: Dublin and Chicago Models," *Urban Studies* 39, no. 10 (2002); Rachel Weber, "Selling City Futures: The Financialization of Urban Redevelopment Policy," *Economic Geography* 86, no. 3 (2010); D. Gibson, "Neighborhood Characteristics and the Targeting of Tax Increment Financing in Chicago," *Journal of Urban Economics* 54, no. 2 (2003).
203. Weber, "Selling City Futures," 258.
204. Weber, "Extracting Value from the City," 536.
205. Andrew Leyshon and Nigel Thrift, "The Capitalization of Almost Everything: The Future of Finance and Capitalism," *Theory Culture & Society* 24, no. 7–8 (2007): 98.
206. Leyshon and Thrift, "Capitalization of Almost Everything," 103.
207. Chris Schwartz and the Neighborhood Capital Budget Group, "NCBG's Chicago TIF Encyclopedia: The First Comprehensive Report on the State of Tax Increment Financing in Chicago" (Chicago: NCBG, 1999).
208. Louik-Schneider and Associates, "Roosevelt-Union Redevelopment Plan and Project, City of Chicago."
209. Louik-Schneider, "Roosevelt-Union Redevelopment Plan," P9.
210. Attributed to Daniel Burnham (1907), quoted in Hall, *Cities of Tomorrow*, 174.
211. Hall, *Cities of Tomorrow*, 181.
212. "Chicago's Next Great Neighborhood," UIC Archives, file RR 44.2.
213. Gene Morgan, "Maxwell Street to Have Face-Lifting Operation," *Chicago Daily News*, May 24, 1939, 17 (where continuation title is given as "Maxwell Street to Have Face Lifted, Ears Scoured").
214. Irene Steyskal, "The Old Order Changeth, Even on Maxwell Street," *Chicago Sunday Tribune*, August 27, 1939, pt. 1, p. 1.
215. On smell and abstraction, see Serres, *Five Senses*.
216. "Maxwell St, to Sell Ethics by the Dozen," *Chicago Tribune*, June 29, 1939, 25.
217. Department of Urban Renewal, "Proposals for Renewal," 8.
218. Department of Urban Renewal, "Proposals for Renewal," 17.
219. James Scott, *Seeing Like a State: How Certain Schemes to Improve the Human Condition Have Failed* (New Haven, CT: Yale University Press, 1998), 2.
220. Scott, *Seeing Like a State*, 329.
221. Wirth, *Ghetto*, 233–34.
222. National Register of Historic Places registration form, 2000, Section 7, 19, Maxwell Street Foundation Archives.
223. UIC University Archives, Associate Chancellor, South Campus Development Records 003/02/02, series 1, box 28, file 28-225 (717 West Maxwell Street

purchase),. Appraisal of 717 West Maxwell Street, Chicago, Illinois—Urban Real Estate Research, Inc. 27.8.92), p. 1.

224. Appraisal of 717 West Maxwell Street by Urban Real Estate Research, Inc., August 27, 1992, UIC University Archives, Associate Chancellor, South Campus Development Records 003/02/02, series 1, box 28, file 28-225 (717 West Maxwell Street purchase), p. 3.

225. Appraisal of 717 West Maxwell Street, 12.

226. Appraisal of 717 West Maxwell Street, 17–18.

227. "Tree Campus USA," https://sustainability.uic.edu/campus-resources/tree-campus-usa/#TreeInventory (accessed June 11, 2016).

228. N. Heynen, H. A. Perkins, and P. Roy, "The Political Ecology of Uneven Urban Green Space: The Impact of Political Economy on Race and Ethnicity in Producing Environmental Inequality in Milwaukee," *Urban Affairs Review* 42, no. 1 (2006).

229. Louik-Schneider, "Roosevelt-Union Redevelopment Plan," 10.

230. Louik-Schneider, "Roosevelt-Union Redevelopment Plan," 12; emphasis in original.

231. Weber, "Extracting Value from the City," 520.

232. Steve Balkin. "Reasons Why the Roosevelt-Union Area Is Ineligible for TIF Designation" (press release for Maxwell Street Preservation Coalition), August 14, 1998.

233. Steve Balkin, "Why the Roosevelt-Union TIF Is Bad Public Policy" (press release for Maxwell Street Preservation Coalition), September 19, 1998.

234. Weber, "Selling City Futures," 258.

235. Chicago Department of Planning, "Future of the Maxwell Street Market," 39–40, UIC Archives, file RR 44.0.

236. See the Maxwell Street Foundation website, http://www.maxwellstreetfoundation.org (accessed May 31, 2016).

237. Tom Smith, "A Visit to Maxwell Street," http://www.chicagobluesguide.com/features/maxwell-street-visit/maxwell-street-visit-page.html (accessed June 11, 2016).

238. Kenneth E. Foote, *Shadowed Ground: America's Landscapes of Violence and Tragedy* (Austin: University of Texas Press, 1997), 5.

239. Ann Swallow, IHPA memorandum to Mayor Richard M. Daley and Charles Thurow, March 9, 1994, Maxwell Street Foundation Archives.

240. Peter Bynoe (Commission on Chicago Landmarks), letter to Ann Swallow (IHPA), May 6, 1994, Maxwell Street Foundation Archives.

241. William Garfield, letter to Ann Swallow (IHPA), May 24, 1994, Maxwell Street Foundation Archives.

242. Lori Grove and Elliot Zashin, letter to IHPA, June 9, 1994, Maxwell Street Foundation Archives.

243. Foote, *Shadowed Ground*; Dolores Hayden, *The Power of Place: Urban Landscapes as Public History* (Cambridge, MA: MIT Press, 1995); Gareth Hoskins, "Materialising Memory at Angel Island Immigration Station, San Francisco," *Environment and Planning A* 32, no. 2 (2007).

244. Hayden, *Power of Place*.

245. William Wheeler (state historic preservation officer), letter to Beth Boland (National Park Service), November 16, 1994, Maxwell Street Foundation Archives.

246. Carol Shull (National Park Service), letter to Lori Grove, December 22, 1994, Maxwell Street Foundation Archives.

247. William Wheeler (state historic preservation officer), letter to Carol Shull (National Park Service), July 13, 2000, Maxwell Street Foundation Archives.

248. Ann Swallow (IHPA), letter to Lori Grove (Maxwell Street Foundation), February 9, 1994. Maxwell Street Foundation Archives.
249. Ann Swallow (IHPA), letter to Mayor Richard M. Daley and Charles Thurow (Chicago Landmarks Commission), March 9, 1994, Maxwell Street Foundation Archives.
250. Plutarch, *Theseus*, trans. John Dryden, http://classics.mit.edu/Plutarch/theseus.html (accessed June 11, 2016).
251. Manuel DeLanda, *A New Philosophy of Society: Assemblage Theory and Social Complexity* (London: Continuum, 2006), 5.
252. National Register of Historic Places registration form, 2000, Section 7, 59.
253. Caitlin Desilvey, "Salvage Memory: Constellating Material Histories on a Hardscrabble Homestead," *Cultural Geographies* 14, no. 3 (2007): 403.
254. "Maxwell Street Heart of Ghetto," *Chicago Daily News*, April 28, 1928, 14.
255. Gene Morgan, "Maxwell Street to Have Face-Lifting Operation," *Chicago Daily News*, May 24, 1939, 1.
256. Lloyd Wendt, "Business Is Always Good on Maxwell Street," *Chicago Sunday Tribune*, October 9, 1947.
257. De Beauvoir, "America Day by Day," 69.
258. Steven Morris, "Maxwell Street Update," *Chicago Tribune*, October 30, 1981. Sec 3, 3.
259. Oscar Katov, "Maxwell St. Has Air of Baghdad Bazaars," *Chicago Sun Times*, May 24, 1951.
260. Terry Cook, "Remembering the Future: Appraisal of Records and the Role of Archives in Constructing Social Memory," in *Archives, Documentation, and Institutions of Social Memory: Essays from the Sawyer Seminar*, ed. Francis X. Blouin and William G. Rosenberg (Ann Arbor: University of Michigan Press, 2006).
261. Carolyn Steedman, *Dust: The Archive and Cultural History* (New Brunswick, NJ: Rutgers University Press, 2002), 10.
262. William J. Turkel, *The Archive of Place: Unearthing the Pasts of the Chilcotin Plateau* (Vancouver: University of British Columbia Press, 2007).
263. Caitlin Desilvey, "Art and Archive: Memory-Work on a Montana Homestead," *Journal of Historical Geography* 33, no. 4 (2007); Charles Withers, "Constructing 'the Geographical Archive,'" *Area* 34, no. 3 (2002); Felix Driver and Lowri Jones, *Hidden Histories of Exploration* (Egham: Royal Holloway, University of London, 2009); Gillian Rose, "Practising Photography: An Archive, a Study, Some Photographers and a Researcher," *Journal of Historical Geography* 26, no. 4 (2000); Matthew Kurtz, "Situating Practices: The Archive and the Filing Cabinet," *Historical Geography* 29 (2001); Kathryn Yusoff, *Bipolar* (London: Arts Catalyst, 2008).
264. Part of catalog heading for Nathan Lerner's Maxwell Street photographs in the Chicago History Museum special collection department.
265. Mike Pearson and Michael Shanks, *Theatre/Archaeology: Disciplinary Dialogues* (London: Routledge, 2001); Hayden Lorimer and Fraser MacDonald, "A Rescue Archaeology, Taransay, Scotland," *Cultural Geographies* 9 (2002); Tim Cresswell and Gareth Hoskins, "Place, Persistence and Practice: Evaluating Historical Significance at Angel Island, San Francisco and Maxwell Street, Chicago," *Annals of the Association of American Geographers* 98, no. 2 (2008); Tim Edensor, *Industrial Ruins: Spaces, Aesthetics, and Materiality* (Oxford: Berg, 2005).
266. Desilvey, "Salvage Memory," 405.
267. Pearson and Shanks, *Theatre/Archaeology*, 156.

268. Turkel, *Archive of Place*; Diana Taylor, *The Archive and the Repertoire: Performing Cultural Memory in the Americas* (Durham, NC: Duke University Press, 2003); Mary Ann Doane, *The Emergence of Cinematic Time: Modernity, Contingency, the Archive* (Cambridge, MA: Harvard University Press, 2002); Miles Ogborn, "Archives," in *Patterned Ground: Entanglements of Nature and Culture*, ed. Stephen Harrison, Steve Pile, and Nigel Thrift (London: Reaktion, 2004); Elizabeth Gagen, Hayden Lorimer, and Alex Vasudevan, eds., *Practicing the Archive: Reflections on Methods and Practice in Historical Geography* (London: Historical Geography Research Group, 2008); Desilvey, "Art and Archive."
269. Ann Laura Stoler, *Along the Archival Grain: Epistemic Anxieties and Colonial Common Sense* (Princeton, NJ: Princeton University Press, 2009), 45.
270. Jacques Derrida, *Archive Fever: A Freudian Impression* (Chicago: University of Chicago Press, 1996); Taylor, *Archive and the Repertoire.*
271. Allan Sekula, "The Body and the Archive," *October*, no. 39 (Winter 1989); Sekula, "Reading an Archive: Photography between Labour and Capital," in *The Photography Reader*, ed. Liz Wells (London: Routedge, 2002).
272. Fragment of catalog entry for "Maxwell Street Photo Collection," University of Illinois Chicago Special Collections.
273. Penelope Papailias, *Genres of Recollection: Archival Poetics and Modern Greece* (New York: Palgrave Macmillan, 2005); Antoinette M. Burton, *Archive Stories: Facts, Fictions, and the Writing of History* (Durham, NC: Duke University Press, 2006).
274. Driver and Jones, *Hidden Histories of Exploration*; Rose, "Practising Photography," 565.
275. Ian Hacking, *Historical Ontology* (Cambridge, MA: Harvard University Press, 2002).
276. Turkel, *Archive of Place*, xix.
277. Text accompanying an image of the Nate's Delicatessen sign, Maxwell Street Foundation website, http://maxwellstreetfoundation.org/sightssounds/nates-delicatessen-sign/ (accessed January 4, 2018).
278. Michael Lynch, "Archives in Formation: Privileged Spaces, Popular Archives and Paper Trails," *History of Human Sciences* 12, no. 2 (1999): 83.
279. Lynch, "Archives in Formation," 67.
280. Steven Connor, "The Necessity of Value," in *Principles Positions*, ed. Judith Squires (London: Lawrence and Wishart, 1993).
281. Connor, "Necessity of Value," 38.
282. Pierre Bourdieu, *Distinction: A Social Critique of the Judgement of Taste* (Cambridge, MA: Harvard University Press, 1984), 45.
283. Gay Hawkins, *The Ethics of Waste: How We Relate to Rubbish* (Lanham, MD: Rowman & Littlefield, 2006), 78.
284. Michael Thompson, *Rubbish Theory: The Creation and Destruction of Value* (Oxford: Oxford University Press, 1979), 9.
285. John Frow, "Invidious Distinction: Waste, Difference, and Classy Stuff," in *Culture and Waste: The Creation and Destruction of Value*, ed. Gay Hawkins and Stephen Muecke (Lanham, MD: Rowman and Littlefield, 2003), 35–36.
286. David Gross, "Objects from the Past," in *Waste Site Stories: The Recycling of Memory*, ed. Brian Neville and Johanne Villeneuve (Albany, NY: SUNY Press, 2002), 31.
287. Yi-fu Tuan, "The Significance of the Artifact," *Geographical Review* 70, no. 4 (1980): 463.

288. Cook, "Remembering the Future," 169.
289. Caption next to wedding photo, Maxwell Street Foundation webpage, http://maxwellstreetfoundation.org/news/know-wedding-photo/ (accessed January 4, 2018).
290. Benjamin, *Arcades Project*, 204–5.
291. Geraldine Pratt, *Working Feminism* (Philadelphia: Temple University Press, 2004), 121.
292. Kathryn Yusoff, "Ice Archives," in Yusoff, *Bipolar*, 120.
293. Desilvey, "Art and Archive," 879.
294. "Buy Much for a Penny," *Chicago Tribune*, September 20, 1896, p. 34.
295. Jack Star, "Maxwell Street Lives," *Chicago Tribune Magazine*, August 24, 1974, 1.
296. Grove, *Chicago's Maxwell Street*.
297. Taylor, *Archive and the Repertoire*.
298. David Atkinson, "Kitsch Geographies and the Everyday Spaces of Social Memory," *Environment and Planning A* 39 (2007); Sam Binkley, "Kitsch as a Repetitive System," *Journal of Material Culture* 5 (2000); Dydia DeLyser, "Collecting, Kitsch and the Intimate Geographies of Social Memory: A Story of Archival Autoethnography," *Transactions of the Institute of British Geographers* 40, no. 2 (2015).
299. Kishik, *Manhattan Project*, 168.
300. Turkel, *Archive of Place*, 66.
301. Binkley, "Kitsch as a Repetitive System," 133.
302. Celeste Olalquiaga, "Holy Kitschen: Collecting Religious Junk from the Street," in *The Object Reader*, ed. Fiona Candlin and Raiford Guins (London: Routledge, 2009), 395.
303. Binkley, "Kitsch as a Repetitive System," 144.
304. Ray Oldenburg, *Celebrating the Third Place: Inspiring Stories About the "Great Good Places" at the Heart of Our Communities* (New York: Marlowe & Co., 2001).
305. Edward Relph, *Place and Placelessness* (London: Pion, 1976), 64.
306. Sharon Zukin, *Naked City: The Death and Life of Authentic Urban Places* (Oxford: Oxford University Press, 2010), xii.
307. www.cityofchicago.org/city/en/depts/dca/supp_info/maxwell_street_market.html (accessed June 30, 2016).

PART THREE

1. Aristotle, *Physics*, bk. 4, in *Complete Works*, 208b34–209a1.
2. Pierre Bourdieu, *Outline of a Theory of Practice* (Cambridge: Cambridge University Press, 1977).
3. Willard Motley Papers, Northern Illinois University Library, Special Collections, box 13, folder 42.
4. Tuan, *Space and Place*, 54.
5. John Agnew, *The United States in the World Economy* (Cambridge: Cambridge University Press, 1987).
6. Municipal Code of Chicago, quoted in Institute of Urban Life, "Diagnostic Survey." 21.
7. Ira Berkow Archives, UIC Special Collections, box 1, file 1-8, p. 8.
8. Benjamin, *Arcades Project*, 88.
9. Institute of Urban Life, "Diagnostic Survey," 3.
10. Raleigh, *Maxwell Street Blues*, 55.

11. Institute of Urban Life, "Diagnostic Survey," 22.
12. Institute of Urban Life, "Diagnostic Survey," 21.
13. Yi-fu Tuan, "Space and Place: Humanistic Perspective," *Progress in Human Geography* 6 (1974): 245.
14. Richard Hartshorne, *The Nature of Geography: A Critical Survey of Current Thought in the Light of the Past* (Lancaster, PA: Association of American Geographers, 1939), 462.
15. Edward Casey, "How to Get from Space to Place in a Fairly Short Stretch of Time," in Feld and Basso, *Senses of Place*, 24.
16. Sharon Wolf, statement at public meeting held by City of Chicago Community Development Commission, October 26, 1993, at YMCA, 1001 West Roosevelt Road, Chicago, UIC University Archives, Associate Chancellor, South Campus Development Records, 003/02/02, series X, box 51, file 432, p. 104.
17. Willard Motley Papers, Northern Illinois University Library, Special Collections, box 13, folder 42, "A Chicagoan Discovers Chicago" (spiralbound notebook with observations of Maxwell Street).
18. Paul C. Adams, Steven D. Hoelscher, and Karen E. Till, *Textures of Place: Exploring Humanist Geographies* (Minneapolis: University of Minnesota Press, 2001), xiii.
19. Robert David Sack, *A Geographical Guide to the Real and the Good* (New York: Routledge, 2003), 41.
20. Robert David Sack, *Homo Geographicus* (Baltimore: Johns Hopkins University Press, 1997), 2–3.
21. On assemblage theory and place, see Colin McFarlane, "The City as Assemblage: Dwelling and Urban Space," *Environment and Planning D: Society and Space* 29, no. 4 (2011); Colin McFarlane and Ben Anderson, "Thinking with Assemblage," *Area* 43, no. 2 (2011); Kim Dovey, *Becoming Places: Urbanism/ Architecture/Identity/Power* (London: Routledge, 2010).
22. DeLanda, *New Philosophy of Society*, 12.
23. DeLanda, *New Philosophy of Society*, 5.
24. DeLanda, *New Philosophy of Society*, 17.
25. Oscar Katov, "Maxwell St. Has Air of Baghdad Bazaars," *Chicago Sun Times*, May 24, 1951.
26. DeLanda, *New Philosophy of Society*, 14.
27. DeLanda, *New Philosophy of Society*, 12.
28. J. D. Dewsbury, "Performativity and the Event: Enacting a Philosophy of Difference," *Environment and Planning D: Society and Space* 18 (2000): 487.
29. DeLanda, *New Philosophy of Society*, 13, 29.
30. Jeff E. Malpas, *Place and Experience: A Philosophical Topography* (Cambridge: Cambridge University Press, 1999), 34.
31. Department of Urban Renewal, "Proposals for Renewal," 12–13.
32. Willard Motley Papers, Northern Illinois University Library, Special Collections, box 13, folder 42.
33. Perec, *Attempt at Exhausting a Place*, 3.
34. Steven Morris, "Maxwell Street Update," *Chicago Tribune*, October 30, 1981, sect. 3, p. 3.
35. Don McKay, *Paradoxides* (Plattsburgh, NJ: McClelland & Stewart, 2012), 55.
36. Kopytoff, "Cultural Biography of Things," 66.
37. Tim Ingold, "Materials against Materiality," *Archaeological Dialogues* 14, no. 1 (2007): 3–4.
38. Feld and Brenneis, "Doing Anthropology in Sound," 469.

39. Matless, *In the Nature of Landscape*, 7.
40. Lloyd Wendt, "Business Is Always Good on Maxwell Street," *Chicago Sunday Tribune*, October 9, 1947.
41. Alan Latham and Derek McCormack, "Moving Cities: Rethinking the Materialities of Urban Geographies," *Progress in Human Geography* 28, no. 6 (2004): 704–5.
42. Yi-Fu Tuan, "Thought and Landscape," in *The Interpretation of Ordinary Landscapes*, ed. Donald Meinig (Oxford: Oxford University Press, 1979), 89.
43. Benjamin, *Arcades Project*, 922.
44. Yi-Fu Tuan, "Language and the Making of Place: A Narrative-Descriptive Approach," *Annals of the Association of American Geographers* 81, no. 4 (1991): 688.
45. Benjamin, *Arcades Project*, 519.
46. "Literally, Maxwell Street No Longer Exists," *Chicago Tribune*, September 20, 1895.
47. Price, *Dry Place*, 4.
48. Kishik, *Manhattan Project*, 45.
49. Perec, *Attempt at Exhausting a Place*, 3.
50. Jacobs, *Death and Life of Great American Cities*; de Certeau, *Practice of Everyday Life*; Seamon, "Body-Subject, Time-Space Routines, and Place-Ballets"; Henri Lefebvre, *Rhythmanalysis: Space, Time, and Everyday Life* (London: Continuum, 2004).
51. Willard Motley Papers, Northern Illinois University Library, Special Collections, box 13, folder 42.
52. Jacobs, *Death and Life of Great American Cities*, 51.
53. Pred, "Place as Historically Contingent Process."
54. Benjamin, *Arcades Project*, 221.
55. Institute of Urban Life, "Diagnostic Survey," 39
56. Serres, *Five Senses*, 279.
57. Martin Heidegger, *Basic Writings: From Being and Time (1927) to the Task of Thinking (1964)* (San Francisco: Harper, 1993), 300.
58. Definition of Maxwell Street Market, Municipal Code of Chicago, quoted in Institute of Urban Life, "Diagnostic Survey," 21.
59. John Berger, *And Our Faces, My Heart, Brief as Photos* (New York: Pantheon, 1984), 56.
60. Rebecca Solnit, *Storming the Gates of Paradise: Landscapes for Politics* (Berkeley: University of California Press, 2007), 1.
61. Lucy Lippard, *The Lure of the Local: Senses of Place in a Multicultural Society* (New York: New Press, 1997), 7.
62. Benjamin, *Arcades Project*, 220.
63. Doreen B. Massey, *Space, Place, and Gender* (Minneapolis: University of Minnesota Press, 1994), 155.
64. Lait and Mortimer, *Chicago: Confidential!*, 60.
65. Doreen Massey, "Landscape as a Provocation: Reflections on Moving Mountains," *Journal of Material Culture* 11, no. 1–2 (2006): 43.
66. Benjamin, *Arcades Project*, 462.
67. Doreen Massey, "Space-Time, Science and the Relationship between Physical Geography and Human Geography," *Transactions of the Institute of British Geographers* 24 (1999): 271.
68. Chicago Department of Planning, "Future of the Maxwell Street Market," 40, UIC Archives, file RR 44.0.
69. Raleigh, *Maxwell Street Blues*, 40.

70. Wirth, *Ghetto*, 240.
71. Stewart, "Precarity's Forms," 519.
72. T. F. Gieryn, "What Buildings Do," *Theory and Society* 31, no. 1 (2002): 35.
73. "Report of a Phase 1 Archaeological Survey of the Maxwell Street Area in Cook County, Illinois," prepared by Archaeological Research, Inc., for the University of Illinois at Chicago, 1994.
74. David Harvey, "The Geography of Capitalist Accumulation," in *Human Geography: An Essential Anthology*, ed. John A. Agnew, David Livingstone, and Alisdair Rogers (1975; Oxford: Blackwell, 1996), 610.
75. Bruno Latour, "On Technical Mediation: Philosophy, Sociology, Genealogy," *Common Knowledge* 4 (1994): 51.
76. Alfred North Whitehead, *The Concept of Nature: The Tarner Lectures Delivered in Trinity College, November 1919* (Mineola, NY: Dover, 2004), 166–67.
77. Ann Swallow (IHPA), letter to Lori Grove, February 9, 1994.
78. Stewart Brand, *How Buildings Learn* (New York: Penguin, 1994), 19.
79. Lloyd Jenkins, "Geography and Architecture: 11 Rue De Conservatoire and the Permeability of Buildings," *Space and Culture* 5, no. 3 (2002): 232.
80. Jane M. Jacobs, "A Geography of Big Things," *Cultural Geographies* 13, no. 1 (2006): 11.
81. Edward S. Casey, *Remembering: A Phenomenological Study* (Bloomington: Indiana University Press, 1987), 186–87.
82. Hayden, *Power of Place*, 46.
83. Mike Royko, "Don't Shed a Tear for Maxwell Street," *Chicago Tribune*, February 11, 1993.
84. Massey, "Space-Time," 274.
85. Price, *Dry Place*, 5.
86. Peggy Phelan, *Unmarked: Politics of Performance* (Routledge, 1993), 146.
87. Judith Butler, "Performativity's Social Magic," in *The Social and Political Body*, ed. T. Schatzki and W. Natter (New York: Guilford, 1996).
88. Bourdieu, *Outline of a Theory of Practice*, 16; de Certeau, *Practice of Everyday Life*.
89. Taylor, *Archive and the Repertoire*, 3.
90. Taylor, *Archive and the Repertoire*, 29.
91. Karen E. Till, *The New Berlin: Memory, Politics, Place* (Minneapolis: University of Minnesota Press, 2005), 14.
92. Taylor, *Archive and the Repertoire*, 20–21.
93. William Garfield, letter to Ann Swallow (IHPA), May 24, 1994.
94. Martin Jones, "Phase Space: Geography, Relational Thinking, and Beyond," *Progress in Human Geography* 33, no. 4 (2009): 497–98.
95. Pred, "Place as Historically Contingent Process," 279.
96. Jones, "Phase Space," 501.
97. DeLanda, *New Philosophy of Society*, 29.
98. Casey, "How to Get from Space to Place," 26.
99. Stephen Lukes, *Power: A Radical View* (London: MacMillan, 1974), 37.
100. Foucault, *Discipline and Punish*, 194.
101. John Allen, *Lost Geographies of Power* (Malden, MA: Blackwell, 2003), 2.
102. "Chicago's Next Great Neighborhood," UIC Archives, file RR 44.2.
103. Robert David Sack, "Human Territoriality: A Theory," *Annals of the Association of American Geographers* 73, no. 1 (1983): 56.
104. Tim Cresswell, *In Place/Out of Place: Geography, Ideology and Transgression* (Minneapolis: University of Minnesota Press, 1996).
105. "Chicago's Next Great Neighborhood," UIC Archives, file RR 44.2.

106. Gerard Kearns and Chris Philo, *Selling Places: The City as Cultural Capital, Past and Present* (Oxford: Pergamon, 1993).

107. "Dining, Retail, Services Area Near UIC Announces Name Change; University Village Maxwell Street Reflects Location's Historic Beginnings," http://www.prweb.com/releases/2015/04/prweb12635107.htm (accessed December 26, 2016).

108. Anna Lowenhaupt Tsing, *Friction: An Ethnography of Global Connection* (Princeton, NJ: Princeton University Press, 2005), 5.

Bibliography

Adams, Paul C., Steven D. Hoelscher, and Karen E. Till. *Textures of Place: Exploring Humanist Geographies*. Minneapolis: University of Minnesota Press, 2001.
Adorno, Theodor, and Walter Benjamin, *Complete Correspondence 1928–1940*. London: Polity, 1999.
Agnew, John. *The United States in the World Economy*. Cambridge: Cambridge University Press, 1987.
Allen, John. *Lost Geographies of Power*. Malden, MA: Blackwell, 2003.
Anderson, Ben, and Paul Harrison. "The Promise of Non-Representational Theories." In *Taking Place: Non-Representational Theories and Geography*, edited by Ben Anderson and Paul Harrison, 1–36. Farnham: Ashgate, 2010.
Anderson, Nels. *The Hobo*. Chicago: University of Chicago Press, 1925.
Appadurai, Arjun. "Introduction: Commodities and the Politics of Value." In *The Social Life of Things: Commodities in Cultural Perspective*, edited by Appadurai, 3–63. Cambridge: Cambridge University Press, 1986.
Aristotle. *The Complete Works of Aristotle: The Revised Oxford Translation*. 2 vols. Bollingen Series. Princeton, NJ: Princeton University Press, 1984.
Atkinson, David. "Kitsch Geographies and the Everyday Spaces of Social Memory." *Environment and Planning A* 39 (2007): 521–40.
Baldwin, James. *Conversations with James Baldwin*. Jackson: University of Mississippi Press, 1989.
Bandt, Ros, Michelle Duffy, and Dolly MacKinnon. *Hearing Places: Sound, Place, Time and Culture*. Newcastle: Cambridge Scholars, 2009.
Barthes, Roland. *Camera Lucida: Reflections on Photography*. Translation of *La chambre claire* (1981). London: Flamingo, 1984.
———. *A Lover's Discourse*. New York: Hill & Wang, 1978.
———. *The Semiotic Challenge*. Berkeley: University of California Press, 1988.
Bauch, Nicholas. *Enchanting the Desert*. Palo Alto, CA: Stanford University Press, 2016.
Bauman, Zygmunt. *Wasted Lives: Modernity and Its Outcasts*. Oxford: Polity, 2003.
Benedetta, Mary. *The Street Markets of London*. London: John Miles, 1936.
Benjamin, Walter. *The Arcades Project*. Cambridge, MA: Belknap Press of Harvard University Press, 1999.
———. "Old Forgotten Children's Books." In *Walter Benjamin: Selected Writings*, vol. 1, *1913–1926*, edited by Marcus Bullock and Michael W. Jennings, 406–13. Cambridge, MA: Harvard University Press, 2004.

———. "A Short History of Photography." In *Classic Essays on Photography*, edited by Alan Trachtenberg, 199–217. New Haven, CT: Leete Island Books, 1980.
Bennett, Jane. *Vibrant Matter: A Political Ecology of Things*. Durham, NC: Duke University Press, 2010.
Berger, John. *And Our Faces, My Heart, Brief as Photos*. 1st ed. New York: Pantheon, 1984.
———. *Ways of Seeing*. Harmondsworth: Penguin, 1972.
Berkow, Ira. *Maxwell Street: Survival in a Bazaar*. Garden City, NY: Doubleday, 1977.
Berman, Marshall. *All That Is Solid Melts into Air: The Experience of Modernity*. 2nd ed. Harmondsworth: Penguin, 1988.
Bhaumik, Subir. *Counter-Gaze: Media, Migrants, Minorities*. Kolkata: Frontpage, 2010.
Bijstervald, Karin. *Mechanical Sound: Technology, Culture, and Public Problems of Noise in the Twentieth Century*. Cambridge, MA: MIT Press, 2008.
Binkley, Sam. "Kitsch as a Repetitive System." *Journal of Material Culture* 5 (2000): 131–52.
Bone, Robert. "Richard Wright and the Chicago Renaissance." *Callaloo* 28 (1986): 446–68.
Booth, Charles. *Life and Labour of the People in London*. 9 vols. London: Macmillan, 1892.
Bourdieu, Pierre. *Distinction: A Social Critique of the Judgement of Taste*. Translated by Richard Nice. Cambridge, MA: Harvard University Press, 1984.
———. *Outline of a Theory of Practice*. Cambridge Studies in Social Anthropology 16. Cambridge: Cambridge University Press, 1977.
———. *Photography: A Middle-Brow Art*. Polity Press, 1990.
Brand, Stewart. *How Buildings Learn*. New York: Penguin, 1994.
Breton, Andre. *Nadja*. 1928; London: Penguin, 1999.
Buck-Morss, Susan. *The Dialectics of Seeing: Walter Benjamin and the Arcades Project*. Cambridge, MA: MIT Press, 1991.
Burgess, Ernest. "The Growth of the City: An Introduction to a Research Project." In *The City: Suggestions for Investigation of Human Behavior in the Urban Environment*, edited by Robert Park and Ernest Burgess, 47–62. Chicago: University of Chicago Press, 1925.
Burton, Antoinette M. *Archive Stories: Facts, Fictions, and the Writing of History*. Durham, NC: Duke University Press, 2006.
Butler, Judith. "Performativity's Social Magic." In *The Social and Political Body*, edited by T. Schatzki and W. Natter, 29–48. New York: Guilford, 1996.
Cahan, Richard, and Michael Williams. *Vivian Maier: Out of the Shadows*. Chicago: CityFiles Press, 2012.
Cappetti, Carla. *Writing Chicago: Modernism, Ethnography and the Novel*. New York: Columbia University Press, 1993.
Casey, Edward S. "How to Get from Space to Place in a Fairly Short Stretch of Time." In *Senses of Place*, edited by Steven Feld and Keith H. Basso, 14–51. Santa Fe, NM: School of American Research, 1996.
———. *Remembering: A Phenomenological Study*. Studies in Phenomenology and Existential Philosophy. Bloomington: Indiana University Press, 1987.
Chatwin, Bruce. *The Songlines*. Harmondsworth: Penguin, 1988.
Chicago Department of City Planning. Development Plan for the Central Area of Chicago: A Definitive Text for Use with Graphic Presentation. Chicago, 1958.
Chicago Department of Urban Renewal. Designation of Slum and Blighted Area—Roosevelt-Halsted. Chicago, 1966.
Chicago Plan Commission. [West Central Area Report.] Chicago, 1951.

Coates, Ta-Nehisi. "The Case for Reparations." *Atlantic*, June 2014. https://www.theatlantic.com/magazine/archive/2014/06/the-case-for-reparations/361631/ (accessed November 5, 2017).

Connor, Steven. "The Necessity of Value." In *Principled Positions*, edited by Judith Squires, 31–49. London: Lawrence and Wishart, 1993.

Cook, Terry. "Remembering the Future: Appraisal of Records and the Role of Archives in Constructing Social Memory." In *Archives, Documentation, and Institutions of Social Memory: Essays from the Sawyer Seminar*, edited by Francis X. Blouin and William G. Rosenberg, 169–81. Ann Arbor: University of Michigan Press, 2006.

Corbin, Alain. *The Foul and the Fragrant: Odor and the French Social Imagination.* Cambridge, MA: Harvard University Press, 1986.

Corner, D. "The Agency of Mapping." In *Mappings*, edited by Denis Cosgrove, 213–52. London: Reaktion, 1999.

Cosgrove, Denis E. "Prospect, Perspective and the Evolution of the Landscape Idea." *Transactions of the Institute of British Geographers* 10 (1985): 45–62.

Crary, Jonathan. *Suspensions of Perception: Attention, Spectacle, and Modern Culture.* Cambridge, MA: MIT Press, 1999.

Cressey, Paul Goalby. *The Taxi-Dance Hall; a Sociological Study in Commercialized Recreation and City Life.* University of Chicago Sociological Series. Chicago: University of Chicago Press, 1932.

Cresswell, Tim. *Fence.* London: Penned in the Margins, 2015.

———. *In Place/out of Place: Geography, Ideology and Transgression.* Minneapolis: University of Minnesota Press, 1996.

———. "Landscape and the Obliteration of Practice." In *Handbook of Cultural Geography*, edited by Kay Anderson, Mona Domosh, Steve Pile, and Nigel Thrift, 269–81. London: Sage, 2003.

———. "The Peninsular of Submerged Hope: Ben Reitman's Social Geography." *Geoforum* 29, no. 2 (May 1998): 207–16.

———. *Soil.* London: Penned in the Margins, 2013.

———. "Towards Topopoetics: Space, Place and the Poem." In *Place, Space and Hermeneutics*, edited by Bruce Janz, 319–331. New York: Springer, 2017.

———. *The Tramp in America.* London: Reaktion, 2001.

Cresswell, Tim, and Gareth Hoskins. "Place, Persistence and Practice: Evaluating Historical Significance at Angel Island, San Francisco and Maxwell Street, Chicago." *Annals of the Association of American Geographers* 98, no. 2 (2008): 392–413.

Crewe, Louise. "Life Itemised: Lists, Loss, Unexpected Significance, and the Enduring Geographies of Discard." *Environment and Planning D: Society and Space* 29, no. 1 (February 2011): 27–46.

Cronon, William. "Kennecott Journey: The Paths Out of Town." In *Under an Open Sky*, edited by Cronon, George Miles and Jay Gitlin, 28–51. New York: Norton, 1992.

———. *Nature's Metropolis: Chicago and the Great West.* New York: Norton, 1991.

———. "A Place for Stories: Nature, History, and Narrative." *Journal of American History* 78, no. 4 (1992): 1347–76.

Darby, H. C. "The Problem of Geographical Description," *Transactions of the Institute of British Geographers* 30 (1962): 1–14.

Darnton, Robert. "Extraordinary Commonplaces." *New York Review of Books*, December 21, 2000, 82–89.

Darrow, Clarence. "Little Louis Epstine," *Pilgrim* 9 (1903). Reprinted in *The Essential Words and Writings of Clarence Darrow*, 172–79. New York: Random House, 2007.

de Beauvoir, Simone. "America Day by Day (L'amerique au jour le jour)." In *Building the Nation: Americans Write about Their Architecture, Their Cities, and Their*

Landscape, edited by Steven Conn and Max Page, 69. Philadelphia: University of Pennsylvania Press, 2003.
de Certeau, Michel. *The Practice of Everyday Life*. Translated by Rendall. Steven. Berkeley: University of California Press, 1984.
DeLanda, Manuel. *A New Philosophy of Society: Assemblage Theory and Social Complexity*. London: Continuum, 2006.
de Leeuw, Sarah. *Geographies of a Lover*. Edmonton: NeWest Press, 2013.
———. *Skeena*. Halfmoon Bay, BC: Caitlin Press, 2015.
Du Bois, W. E. B. *The Souls of Black Folk*. New York: Dover, 1903.
DeLyser, Dydia. "Collecting, Kitsch and the Intimate Geographies of Social Memory: A Story of Archival Autoethnography." *Transactions of the Institute of British Geographers* 40, no. 2 (April 2015): 209–22.
Department of Urban Renewal "Roosevelt-Halsted: Proposals for Renewal." Chicago, 1966.
Derrida, Jacques. *Archive Fever: A Freudian Impression*. Chicago: University of Chicago Press, 1996.
———. *Copy, Archive, Signature: A Conversation on Photography*. Stanford, CA: Stanford University Press, 2010.
Desilvey, Caitlin. "Art and Archive: Memory-Work on a Montana Homestead." *Journal of Historical Geography* 33, no. 4 (2007): 878–900.
———. "Salvage Memory: Constellating Material Histories on a Hardscrabble Homestead." *Cultural Geographies* 14, no. 3 (2007): 401–24.
Dewsbury, J. D. "Language and the Event: The Unthought of Appearing Worlds." In *Taking Place: Non-Representational Theories and Geography*, edited by Ben Anderson and Paul Harrison, 147–60. Farnham: Ashgate, 2010.
———. "Performativity and the Event: Enacting a Philosophy of Difference." *Environment and Planning D: Society and Space* 18 (2000): 473–96.
Doane, Mary Ann. *The Emergence of Cinematic Time: Modernity, Contingency, the Archive*. Cambridge, MA: Harvard University Press, 2002.
Douglas, Mary. *Purity and Danger: An Analysis of Concepts of Pollution and Taboo*. New York: Praeger, 1966.
Dovey, Kim. *Becoming Places: Urbanism/Architecture/Identity/Power*. London: Routledge, 2010.
Driver, Felix. "Moral Geographies: Social Science and the Urban Environment in Mid-Nineteenth Century England." *Transactions of the Institute of British Geographers* 13 (1988): 275–87.
———. *Power and Pauperism: The Workhouse System, 1834–1884*. Cambridge: Cambridge University Press, 1993.
Driver, Felix, and Lowri Jones. *Hidden Histories of Exploration*. Egham: Royal Holloway, University of London, 2009.
Dyer, Geoff. *The Ongoing Moment*. London: Little, Brown, 2005.
Eco, Umberto. *The Infinity of Lists*. London: MacLehose, 2009.
Edensor, Tim. *Industrial Ruins: Spaces, Aesthetics, and Materiality*. Oxford: Berg, 2005.
Edwards, Elizabeth. *Anthropology and Photography, 1860–1920*. New Haven, CT: Yale University Press, in association with the Royal Anthropological Institute London, 1992.
Emerson, Ralph Waldo. *The Conduct of Life*. Boston: Ticknor and Fields, 1860.
Eshel, Shuli, and Roger Schatz. *Jewish Maxwell Street Stories*. Voices of America. Charleston, SC: Arcadia, 2004.
Feld, Steven. "Waterfalls of Song: An Acoustemology of Place Resounding in Bosavi, Papua New Guinea." In *Senses of Place*, edited by Feld and Keith H. Basso, 91–136. Santa Fe, NM: School of American Research Press, 1996.

Feld, Steven, and Donald Brenneis. "Doing Anthropology in Sound." *American Ethnologist* 31, no. 4 (November 2004): 461–74.
Feng, Peter X. *Screening Asian Americans*. Rutgers Depth of Field Series. New Brunswick, NJ: Rutgers University Press, 2002.
Fish, Stanley. *How to Write a Sentence*. New York: HarperCollins, 2011.
———. "Barack Obama's Prose Style." *New York Times*, January 22, 2009 (https://fish.blogs.nytimes.com/2009/01/22/barack-obamas-prose-style/?scp=3&sq=stanley%20fish&st=cse)
Foote, Kenneth E. *Shadowed Ground: America's Landscapes of Violence and Tragedy*. 1st ed. Austin: University of Texas Press, 1997.
Foucault, Michel. *Discipline and Punish: The Birth of the Prison*. New York: Vintage, 1979.
———. *The Order of Things: An Archaeology of the Human Sciences*. 1st American ed. New York: Pantheon, 1971.
Frank, Robert. *The Americans: Photographs*. New York: Grove, 1959.
Frow, John. "Invidious Distinction: Waste, Difference, and Classy Stuff." In *Culture and Waste: The Creation and Destruction of Value*, edited by Gay Hawkins and Stephen Muecke, 25–38. Lanham, MD: Rowman and Littlefield, 2003.
Gagen, Elizabeth, Hayden Lorimer, and Alex Vasudevan, eds. *Practicing the Archive: Reflections on Methods and Practice in Historical Geography*. Historical Geography Research Group Monograph Series. London: Historical Geography Research Group, 2008.
Geertz, Clifford. *The Interpretation of Cultures*. New York: Basic Books, 1973.
Gibson, D. "Neighborhood Characteristics and the Targeting of Tax Increment Financing in Chicago." *Journal of Urban Economics* 54, no. 2 (September 2003): 309–27.
Gieryn, T. F. "What Buildings Do." *Theory and Society* 31, no. 1 (2002): 35–74.
Gilbert, James Burkhart. *Perfect Cities: Chicago's Utopias of 1893*. Chicago: University of Chicago Press, 1991.
Giles, James R., and Jerome Klinkowitz. "The Emergence of Willard Motley in Black American Literature." *Negro American Literature Forum* 6, no. 2 (1972): 31–34.
Goldensohn, Lorrie. *Elizabeth Bishop: The Biography of a Poetry*. New York: Columbia University Press, 1992.
Goldsmith, Kenneth. *Capital: New York, Capital of the 20th Century*. New York: Verso, 2015.
———. "Rewriting Walter Benjamin's 'The Arcades Project.'" http://www.poetryfoundation.org/harriet/2011/04/rewriting-walter-benjamins-the-arcades-project (accessed February 1, 2014).
Gordon, Colin. "Blighting the Way: Urban Renewal, Economic Development, and the Elusive Definition of Blight." *Fordham Urban Law Journal* 31, no. 2 (2003): 305–37.
Gross, David. "Objects from the Past." In *Waste Site Stories: The Recycling of Memory*, edited by Brian Neville and Johanne Villeneuve, 29–37. Albany, NY: SUNY Press, 2002.
Grove, Lori. *Chicago's Maxwell Street*. Images of America. Chicago: Aracadia, 2002.
Hacking, Ian. *Historical Ontology*. Cambridge, MA: Harvard University Press, 2002.
Hall, Peter Geoffrey. *Cities of Tomorrow: An Intellectual History of Urban Planning and Design in the Twentieth Century*. 3rd ed. 1988; Oxford: Blackwell, 2002.
Hannerz, Ulf. *Exploring the City: Inquiries toward an Urban Anthropology*. New York: Columbia University Press, 1980.
Harley, J. B. "Deconstructing the Map." *Cartographica* 26 (1989): 1–20.
Harper, Douglas A. *Visual Sociology*. New York: Routledge, 2012.

Hart, John Fraser. "The Highest Form of the Geographer's Art." *Annals of the Association of American Geographers* 72, no. 1 (1981): 1–29.
Hartshorne, Richard. *The Nature of Geography: A Critical Survey of Current Thought in the Light of the Past*. Lancaster, PA: Association of American Geographers, 1939.
Harvey, David. "The Geography of Capitalist Accumulation." In *Human Geography: An Essential Anthology*, edited by John A. Agnew, David Livingstone, and Alisdair Rogers, 600–622. 1975; Oxford: Blackwell, 1996.
———. *The Limits to Capital*. Oxford B. Blackwell, 1982.
———. "Neo-Liberalism as Creative Destruction." *Geografiska Annaler Series B: Human Geography* 88B, no. 2 (2006): 145–58.
———. "The Right to the City." *New Left Review* 11 (2008): 23–40.
Hauser, Susanne. "Waste into Heritage: Remarks on Materials in the Arts, on Memories and the Museum." In *Waste Site Stories: The Recycling of Memory*, edited by Brian Neville and Johanne Villeneuve, 39–54. Albany, NY: SUNY Press, 2002.
Hawkins, Gay. *The Ethics of Waste: How We Relate to Rubbish*. Lanham, MD: Rowman & Littlefield, 2006.
Hawkins, Harriet. *For Creative Geographies: Geography, Visual Arts and the Making of Worlds*. London: Routledge, 2013.
Hayden, Dolores. *The Power of Place: Urban Landscapes as Public History*. Cambridge, MA: MIT Press, 1995.
Heap, Chad. *Slumming: Sexual and Racial Encounters in American Nightlife, 1885–1940*. Chicago: University of Chicago Press, 2009.
Heat-Moon, William Least. *Prairyerth (a Deep Map)*. Boston: Houghton Mifflin, 1991.
Heathcott, Joseph, and Angela Dietz, *Capturing the City: Photographs from the Streets of St. Louis, 1900–1930*. St. Louis: Missouri History Museum Press, 2016.
Heidegger, Martin. *Basic Writings: From Being and Time (1927) to the Task of Thinking (1964)*. Rev. and expanded ed. San Francisco: Harper, 1993.
Heynen, N., H. A. Perkins, and P. Roy. "The Political Ecology of Uneven Urban Green Space: The Impact of Political Economy on Race and Ethnicity in Producing Environmental Inequality in Milwaukee." *Urban Affairs Review* 42, no. 1 (September 2006): 3–25.
Highmore, Ben. *Everyday Life and Cultural Theory: An Introduction*. London: Routledge, 2002.
Hoskins, Gareth. "Materialising Memory at Angel Island Immigration Station, San Francisco." *Environment and Planning A* 32, no. 2 (2007): 437–55.
Howe, Susan. *Pierce-Arrow*. New York: New Directions, 1999.
Ingold, Tim. "Materials against Materiality." *Archaeological Dialogues* 14, no. 1 (2007): 1–16.
Institute of Urban Life, Loyola University. "Diagnostic Survey of Relocation Problems of Non-Residential Establishments, Roosevelt-Halsted Area." Prepared for the Department of Urban Renewal, City of Chicago, 1965.
Jackson, Peter. "Constructions of Culture, Representations of Race; Edward Curtis's 'Way of Seeing.'" In *Inventing Places: Studies in Cultural Geography*, edited by Kay Anderson and Fay Gale, 89–106. Melbourne: Longman, 1992.
———. "Social Disorganization and Moral Order in the City." *Transactions of the Institute of British Geographers* 9 (1984): 168–80.
Jacobs, Jane. *The Death and Life of Great American Cities*. New York: Vintage, 1961.
———. *The Economy of Cities*. New York: Cape, 1969.
Jacobs, Jane M. "A Geography of Big Things." *Cultural Geographies* 13, no. 1 (2006): 1–27.
Jagodzinski, Jan, ed. *Psychoanalyzing Cinema: A Productive Encounter with Lacan, Deleuze, and Žižek*. New York: Palgrave Macmillan, 2012.

Jamieson, M. "The Place of Counterfeits in Regimes of Value: An Anthropological Approach." *Journal of the Royal Anthropological Institute* 5, no. 1 (Mar 1999): 1–11.
Jenkins, Lloyd. "Geography and Architecture: 11 Rue De Conservatoire and the Permeability of Buildings." *Space and Culture* 5, no. 3 (2002): 222–36.
Jones, Martin. "Phase Space: Geography, Relational Thinking, and Beyond." *Progress in Human Geography* 33, no. 4 (2009): 487–506.
Kaiser, Ward L., and Denis Wood. *Seeing through Maps: The Power of Images to Shape Our World View*. Amherst, MA: ODT Inc., 2001.
Kearns, Gerard, and Chris Philo, eds. *Selling Places: The City as Cultural Capital, Past and Present*. Policy, Planning, and Critical Theory. Oxford: Pergamon, 1993.
Kishik, David. *The Manhattan Project: A Theory of a City*. Palo Alto, CA: Stanford University Press, 2015.
Kitagawa, Evelyn Mae, and Karl E. Taeuber, eds. *Local Community Fact Book: Chicago Metropolitan Area, 1960*. Chicago Community Inventory, University of Chicago, 1963.
Kopytoff, Igor. "The Cultural Biography of Things: Commoditization as Process." In *The Social Life of Things: Commodities in Cultural Perspective*, edited by Arjun Appadurai, 64–94. Cambridge: Cambridge University Press, 1986.
Kurtz, Matthew. "Situating Practices: The Archive and the Filing Cabinet." *Historical Geography* 29 (2001): 26–37.
Lait, Jack, and Lee Mortimer. *Chicago: Confidential!* New York: Crown, 1950.
———. *New York: Confidential!* Chicago: Ziff-Davis, 1948.
———. *Washington: Confidential!* New York: Crown, 1951.
Latham, Alan, and Derek McCormack. "Moving Cities: Rethinking the Materialities of Urban Geographies." *Progress in Human Geography* 28, no. 6 (2004): 701–24.
Latour, Bruno. "On Technical Mediation: Philosophy, Sociology, Genealogy." *Common Knowledge* 4 (1994): 29–64.
Lefebvre, Henri. *Rhythmanalysis: Space, Time, and Everyday Life*. London: Continuum, 2004.
Lerner, Nathan, and Jo-Ann Conklin. *Nathan Lerner's Maxwell Street*. Exhibition catalog. Iowa City: University of Iowa Museum of Art, 1993.
Leyshon, Andrew, and Nigel Thrift. "The Capitalization of Almost Everything: The Future of Finance and Capitalism." *Theory Culture & Society* 24, no. 7–8 (December 2007): 97–115.
Lichtenstein, Therese. "The City in Twilight." In *Twilight Visions: Surrealism and Paris*, edited by Lichtenstein, 11–70. Berkeley: University of California Press, 2009.
Lindner, Rolf. *The Reportage of Urban Culture: Robert Park and the Chicago School*. Cambridge: Cambridge University Press, 1996.
Lippard, Lucy. *The Lure of the Local: Senses of Place in a Multicultural Society*. New York: New Press, 1997.
Lorimer, Hayden, and Fraser MacDonald. "A Rescue Archaeology, Taransay, Scotland." *Cultural Geographies* 9 (2002): 95–103.
Lukes, Stephen. *Power: A Radical View*. London: MacMillan, 1974.
Lynch, Michael. "Archives in Formation: Privileged Spaces, Popular Archives and Paper Trails." *History of Human Sciences* 12, no. 2 (1999): 65–87.
Madden, Raymond. *Being Ethnographic: A Guide to the Theory and Practice of Ethnography*. Thousand Oaks, CA: SAGE Publications, 2010.
Magrane, Eric and Christopher Cokinos eds. *The Sonoran Desert: A Literary Field Guide*. Tucson: University of Arizona Press, 2016.
Maloof, John. *Vivian Maier: Street Photographer*. Brooklyn, NY: PowerHouse Books, 2011.

Malpas, Jeff E. *Heidegger's Topology: Being, Place, World.* Cambridge, MA: MIT Press, 2008.
———. *Place and Experience: A Philosophical Topography.* Cambridge: Cambridge University Press, 1999.
Massey, Doreen. "Landscape as a Provocation: Reflections on Moving Mountains." *Journal of Material Culture* 11, no. 1–2 (March–July 2006): 33–48.
———. *Space, Place, and Gender.* Minneapolis: University of Minnesota Press, 1994.
———. "Space-Time, Science and the Relationship between Physical Geography and Human Geography." *Transactions of the Institute of British Geographers* 24 (1999): 261–79.
Matless, David. *In the Nature of Landscape: Cultural Geography on the Norfolk Broads.* Oxford: Wiley Blackwell, 2014.
Mayhew, Henry. *Mayhew's London (Selections from London Labour and London Poor).* London: Spring Books, 1851.
McFarlane, Colin. "The City as Assemblage: Dwelling and Urban Space." *Environment and Planning D: Society and Space* 29, no. 4 (August 2011): 649–71.
McFarlane, Colin, and Ben Anderson. "Thinking with Assemblage." *Area* 43, no. 2 (June 2011): 162–64.
McGreal, S., J. Berry, G. Lloyd, and J. McCarthy. "Tax-Based Mechanisms in Urban Regeneration: Dublin and Chicago Models." *Urban Studies* 39, no. 10 (September 2002): 1819–31.
McGuire, Richard. *Here.* New York: Pantheon, 2014.
McKay, Don. *Paradoxides.* Plattsburgh, NJ: McClelland & Stewart, 2012.
Meinig, Donald. "Geography as an Art," *Transactions of the Institute of British Geographers* 8 (1983): 314–28.
Merriman, P., G. Revill, T. Cresswell, H. Lorimer, D. Matless, G. Rose, and J. Wylie. "Landscape, Mobility, Practice." *Social & Cultural Geography* 9, no. 2 (2008): 191–212.
Miller, William Ian. *The Anatomy of Disgust.* Cambridge, MA: Harvard University Press, 1997.
Morales, Alfonso. "Making Money at the Market: The Social and Economic Logic of Informal Markets." PhD diss., Northwestern University, 1993.
Motley, Willard. *Knock on Any Door.* New York: Appleton-Century, 1947; DeKalb: Northern Illinois University Press, 2001.
———. *Let No Man Write My Epitaph.* New York: Random House, 1958.
Motley, Willard, and Jerome Klinkowitz. *The Diaries of Willard Motley.* Ames: Iowa State University Press, 1979.
Mulvey, Laura. "Visual Pleasure and Narrative Cinema." *Screen* 16, no. 3 (1975): 6–18.
Nash, Catherine. "Reclaiming Vision: Looking at Landscape and the Body." *Gender, Place and Culture* 3, no. 2 (1996): 149–69.
Nelson, Maggie. *The Argonauts.* Seattle: Wave Books, 2015.
———. *Bluets.* Seattle: Wave Books, 2009.
———. *Jane: A Murder.* Brooklyn, NY: Soft Skull, 2005.
Ogborn, Miles "Archives." In *Patterned Ground: Entanglements of Nature and Culture,* edited by Stephen Harrison, Steve Pile, and Nigel Thrift, 240–42. London: Reaktion, 2004.
Ogden, Laura. *Swamplife: People, Gators, and Mangroves Entangled in the Everglades.* Minneapolis: University of Minnesota Press, 2011.
Olalquiaga, Celeste. "Holy Kitschen: Collecting Religious Junk from the Street." In *The Object Reader,* edited by Fiona Candlin and Raiford Guins, 391–405. London: Routledge, 2009.

Oldenburg, Ray. *Celebrating the Third Place: Inspiring Stories About the "Great Good Places" at the Heart of Our Communities*. New York: Marlowe & Co., 2001.

Papailias, Penelope. *Genres of Recollection: Archival Poetics and Modern Greece*. New York: Palgrave Macmillan, 2005.

Park, Robert, and Ernest Burgess. *The City: Suggestions for Investigation of Human Behavior in the Urban Environment*. Chicago: University of Chicago Press, 1925.

Pearson, Mike, and Michael Shanks. *Theatre/Archaeology: Disciplinary Dialogues*. London: Routledge, 2001.

Perec, Georges. *An Attempt at Exhausting a Place in Paris*. Imagining Science. Cambridge, MA: Wakefield Press, 2010.

———. *Species of Spaces and Other Pieces*. New York: Penguin, 1997.

Perelman, Bob. "Parataxis and Narrative: The New Sentence in Theory and Practice." *American Literature* 65, no. 2 (1993): 313–24.

Phelan, Peggy. *Unmarked: Politics of Performance*. New York: Routledge, 1993.

Philo, Chris. "The Child in the City." *Journal of Rural Studies* 8, no. 2 (1992): 193–207.

Polacheck, Hilda Satt. *I Came a Stranger: The Story of a Hull-House Girl*. Urbana-Champaign: University of Illinois Press, 1991.

Pratt, Geraldine. *Working Feminism*. Philadelphia: Temple University Press, 2004.

Pred, Allan Richard. "Hypermodernity, Identity and the Montage Form." In *Space and Social Theory: Interpreting Modernity and Postmodernity*, edited by Georges Benko and Ulf Strohmayer, 119–40. Oxford: Blackwell, 1997.

———. "Place as Historically Contingent Process: Structuration and the Time-Geography of Becoming Places." *Annals of the Association of American Geographers* 74, no. 2 (1984): 279–97.

Price, Patricia L. *Dry Place: Landscapes of Belonging and Exclusion*. Minneapolis: University of Minnesota Press, 2004.

Pritchett, Wendell E. "The 'Public Menace' of Blight: Urban Renewal and the Private Uses of Eminent Domain." *Yale Law and Policy Review* 21, no. 2 (2003): 1–52.

Prokopoff, Steven. *Nathan Lerner: A Photographic Retrospective 1932–1979*. Exhibition catalog. Boston: Institute of Contemporary Art, 1979.

Raleigh, Michael. *The Maxwell Street Blues: A Chicago Mystery Featuring Paul Whelan*. New York: St. Martin's Press, 1994.

Rankine, Claudia. *Citizen: An American Lyric*. Minneapolis, MN: Graywolf, 2014.

Relph, Edward. *Place and Placelessness*. London: Pion, 1976.

Rendell, Jane. *Site-Writing: The Architecture of Art Criticism*. London: I. B. Tauris, 2010.

Riis, Jacob A. *How the Other Half Lives*. New York: Charles Scribner's Sons, 1890.

———. *The Making of an American*. New York: Macmillan, 1901.

Rodaway, Paul. *Sensuous Geographies: Body, Sense, and Place*. London: Routledge, 1994.

Rose, Gillian. "Engendering the Slum: Photography in East London in the 1930s." *Gender Place and Culture* 4, no. 3 (1997): 277–300.

———. "Practising Photography: An Archive, a Study, Some Photographers and a Researcher." *Journal of Historical Geography* 26, no. 4 (2000): 555–71.

Rosler, Martha. "In, around, and Afterthoughts (on Documentary Photography)." In *The Contest of Meaning: Critical Histories of Photography*, edited by Richard Bolton, 303–33. Cambridge, MA: MIT Press, 1989.

Rossi, Peter H., and Robert A. Dentler. *The Politics of Urban Renewal: The Chicago Findings*. New York: Free Press of Glencoe, 1961.

Rothstein, Richard. *The Color of Law: A Forgotten History of How Our Government Segregated America*. New York: Liveright, 2017.

Ryan, James R. *Picturing Empire: Photography and the Visualization of the British Empire*. London: Reaktion, 1997.
Sack, Robert David. *Homo Geographicus*. Baltimore: Johns Hopkins University Press, 1997.
———. *A Geographical Guide to the Real and the Good*. New York: Routledge, 2003.
———. "Human Territoriality: A Theory." *Annals of the Association of American Geographers* 73, no. 1 (1983): 55–74.
Said, Edward W. *Orientalism*. New York: Vintage, 1979.
Schlosser, K. "Regimes of Ethical Value? Landscape, Race and Representation in the Canadian Diamond Industry." *Antipode* 45, no. 1 (January 2013): 161–79.
Schulz, Edward C. "A Functional Analysis of Retail Trade in the Maxwell Street Market Area of Chicago." Master's thesis, Northwestern University, 1954.
Schwartz, Joan M., and James M. Ryan. *Picturing Place: Photography and the Geographical Imagination*. London: I. B. Tauris, 2003.
Scott, Clive. *Street Photography: From Atget to Cartier-Bresson*. London: I. B. Tauris, 2007.
Scott, James. *Seeing Like a State: How Certain Schemes to Improve the Human Condition Have Failed*. New Haven, CT: Yale University Press, 1998.
Seamon, David. "Body-Subject, Time-Space Routines, and Place-Ballets." In *The Human Experience of Space and Place*, edited by Anne Buttimer and David Seamon, 148–65. London: Croom Helm, 1980.
Sekula, Allan. "The Body and the Archive." *October*, no. 39 (Winter 1989), 65–108.
———. "On the Invention of Photographic Meaning." In *Photography in Print: Writings from 1816 to the Present*, edited by Vicki Goldberg, 452–73. Albuquerque: University of New Mexico Press, 1981.
———. "Reading an Archive: Photography between Labour and Capital." In *The Photography Reader*, edited by Liz Wells, 443–52. London: Routedge, 2002.
Serres, Michel. *The Five Senses: A Philosophy of Mingled Bodies*. London: Continuum, 2008.
Shaw, Clifford R., and E. W. Burgess. *The Jack-Roller: A Delinquent Boy's Own Story*. Chicago: University of Chicago Press, 1930.
Sibley, David. "Gender, Science, Politics and Geographies of the City." *Gender, Place and Culture* 2, no. 1 (1995): 37–50.
Smith, Laurel. "Chips Off the Old Ice Block: *Nanook of the North* and the Relocation of Cultural Identity." In *Engaging Film: Geographies of Identity and Mobility*, edited by Tim Cresswell and Deborah Dixon, 94–122. London: Rowman and Littlefield, 2001.
Smith, Shawn Michelle. *American Archives: Gender, Race, and Class in Visual Culture*. Princeton, NJ: Princeton University Press, 2000.
Solnit, Rebecca. *A Field Guide to Getting Lost*. London: Penguin, 2006.
———. *Storming the Gates of Paradise: Landscapes for Politics*. Berkeley: University of California Press, 2007.
Sontag, Susan. *On Photography*. New York: Farrar, Straus and Giroux, 1977.
Springer, Simon, "Earth Writing," *GeoHumanities* 3, no 1 (2017): 1–19.
Stallybrass, Peter, and Allon White. *The Politics and Poetics of Transgression*. Ithaca, NY: Cornell University Press, 1986.
Steedman, Carolyn. *Dust: The Archive and Cultural History*. New Brunswick, NJ: Rutgers University Press, 2002.
Stewart, Kathleen. *Ordinary Affects*. Durham, NC: Duke University Press, 2007.
———. "Precarity's Forms." *Cultural Anthropology* 27, no. 3 (2012): 518–25.
———. *A Space on the Side of the Road*. Princeton, NJ: Princeton University Press, 1996.
Stoler, Ann Laura. *Along the Archival Grain: Epistemic Anxieties and Colonial Common Sense*. Princeton, NJ: Princeton University Press, 2009.

Strasser, Susan. *Waste and Want: A Social History of Trash*. New York: Holt, 2000.
Straw, Will. "Urban Confidential: The Lurid City of the 1950s." In *The Cinematic City*, edited by David Clarke, 110–28. London: Routledge, 1997.
Taylor, Diana. *The Archive and the Repertoire: Performing Cultural Memory in the Americas*. Durham, NC: Duke University Press, 2003.
Thompson, Michael. *Rubbish Theory: The Creation and Destruction of Value*. Oxford: Oxford University Press, 1979.
Thomson, J., and Adolphe Smith. *Street Life in London. By J. Thomson . . . and Adolphe Smith. With Permanent Photographic Illustrations Taken from Life Expressly for This Publication*. London: Sampson Low, Marston, Searle, & Rivington, 1877.
Thrasher, Frederic Milton. *The Gang: A Study of 1,313 Gangs in Chicago*. 2nd ed. Chicago: University of Chicago Press, 1936.
Thrift, Nigel. "All Nose." In *Handbook of Cultural Geography*, edited by Kay Anderson, Mona Domosh, Steve Pile, and Nigel Thrift, 9–13. London: Sage, 2002.
Thrift, Nigel., and J. D. Dewsbury. "Dead Geographies—and How to Make Them Live." *Environment and Planning D: Society and Space* 18, no. 4 (August 2000): 411–32.
Till, Karen E. *The New Berlin: Memory, Politics, Place*. Minneapolis: University of Minnesota Press, 2005.
Trotter, David. "The New Historicism and the Psychopathology of Everyday Modern Life." In *Filth: Dirt, Disgust, and Modern Life*, 30–48. Minneapolis: University of Minnesota Press, 2005.
Tsing, Anna Lowenhaupt. *Friction: An Ethnography of Global Connection*. Princeton, NJ: Princeton University Press, 2005.
———. *The Mushroom at the End of the World: On the Possibility of Life in Capitalist Ruins*. Princeton, NJ: Princeton University Press, 2015.
Tuan, Yi-Fu. "Language and the Making of Place: A Narrative-Descriptive Approach." *Annals of the Association of American Geographers* 81, no. 4 (1991): 684–96.
———. "The Significance of the Artifact." *Geographical Review* 70, no. 4 (1980): 462–72.
———. "Space and Place: Humanistic Perspective." *Progress in Human Geography* 6 (1974): 211–52.
———. *Space and Place: The Perspective of Experience*. Minneapolis: University of Minnesota Press, 1977.
———. "Thought and Landscape." In *The Interpretation of Ordinary Landscapes*, edited by Donald Meinig, 89–102. Oxford: Oxford University Press, 1979.
Tucker, Jennifer. "Eye on the Street: Photography in Urban Public Spaces." *Radical History Review*, no. 114 (Fall 2012), 7–18.
Turkel, William J. *The Archive of Place: Unearthing the Pasts of the Chilcotin Plateau*. Nature, History, Society. Vancouver: University of British Columbia Press, 2007.
Urry, John. *The Tourist Gaze*. 2nd ed. London: Sage, 2001.
Valentine, Gill. "Children Should Be Seen and Not Heard: The Production and Transgression of Adult's Public Space." *Urban Geography* 17, no. 3 (1996): 205–20.
Veiller, Lawrence. "Slum Clearance." In *Housing Problems in America*, proceedings of the tenth National Housing Association conference, Philadelphia, January 28–30, 1929, 71–84. New York: National Housing Association, 1929.
Ward, Colin. *The Child in the City*. London: Architectural Press, 1978.
Ward, David. *Poverty, Ethnicity and the American City, 1840–1925*. Cambridge: Cambridge University Press, 1989.
Ware, Chris. *Building Stories*. New York: Pantheon, 2012.
———. "Chris Ware on *Here* by Richard McGuire—a Game-Changing Graphic Novel." *Guardian*, December 17, 2014. https://www.theguardian.com/books/2014/dec/17/chris-ware-here-richard-mcguire-review-graphic-novel
Weber, Max. *The City*. London: Heinemann, 1960.

———. "The Nature of the City." In *Classic Essays on the Culture of Cities*, edited by Richard Sennett, 23–46. Englewood Cliffs, NJ: Prentice-Hall, Inc., 1969.
Weber, Rachel. "Extracting Value from the City: Neoliberalism and Urban Redevelopment." *Antipode* 34, no. 3 (June 2002): 519–40.
———. "Selling City Futures: The Financialization of Urban Redevelopment Policy." *Economic Geography* 86, no. 3 (July 2010): 251–74.
Westerbeck, Colin, and Joel Meyerowitz. *Bystander: A History of Street Photography, with an Afterword on Street Photography since the 1970s*. Boston: Bulfinch, 2001.
White, Hayden. "The Question of Narrative in Contemporary Historical Theory." *History and Theory* 23, no. 1 (1984): 1–33.
Whitehead, Alfred North. *The Concept of Nature: The Tarner Lectures Delivered in Trinity College, November 1919*. Mineola, NY: Dover, 2004.
Wirth, Louis. *The Ghetto*. Chicago: University of Chicago Press, 1928.
Withers, Charles. "Constructing 'the Geographical Archive.'" *Area* 34, no. 3 (2002): 303–11.
Yusoff, Kathryn. *Bipolar*. London: Arts Catalyst, 2008.
———. "Ice Archives." In *Bipolar*, edited by Yusoff, 116–23. London: Arts Catalyst, 2008.
Zeublin, Charles. "The Chicago Ghetto." In *Hull House Maps and Papers*, edited by Jane Addams, 91–111. New York: Thomas Y. Crowell & Co., 1895.
Zipp, Samuel. *Manhattan Projects: The Rise and Fall of Urban Renewal in Cold War New York*. Oxford: Oxford University Press, 2010.
Zorbaugh, Harvey Warren. *The Gold Coast and the Slum: A Sociological Study of Chicago's Near North Side*. Chicago: University of Chicago Press, 1929.
Zukin, Sharon. *Naked City: The Death and Life of Authentic Urban Places*. Oxford: Oxford University Press, 2010.

Index